Agricultural Pollution

Spon's Environmental Science and Engineering Series

This new series covers a wide range of water, waste and contaminated land issues in the context of current best practice, perception and legislation.

Series coverage is broad. On the water and wastewater side it includes resource management, treatment, distribution and collection, monitoring and regulation. In relation to waste management it extends across the waste hierarchy, covering a range of technical and regulatory issues in areas such as waste minimisation, separation and sorting, recycling and disposal, and the determination of appropriate waste strategy.

The series is targeted at engineers and scientists in the process, waste, and environmental sectors. Titles will also be of interest to economists, lawyers, legislators, regulators, and advanced students.

Series Editor
Jeremy Joseph, JBJ Environment, Hydrogeological and Waste Management Consultant, C/o Spon Press, 11 New Fetter Lane, London EC4P 4EE, UK

Topics under consideration for the series include

- Potable/usable water
- Waste and polluted waters and sludges
- Water control and management issues
- Contaminated land
- Non-disposal strategies for waste
- Waste disposal routes
- Wastes from agriculture

The series editor will be pleased to hear from potential authors interested in writing on any topics relevant to the series including, but not limited to, the issues cited above. Please contact him at the above address with an indication of the scope of any proposed volume together with details regarding its intended readership.

All volumes are published by Spon Press, part of the Taylor and Francis Group, and are sold through our worldwide distribution networks.

Visit us on the web at:
www.sponpress.com

Agricultural Pollution

Environmental problems and practical solutions

Graham Merrington, Linton Winder,
Robert Parkinson and Mark Redman

London and New York

First published 2002
by Spon Press
11 New Fetter Lane, London EC4P 4EE

Simultaneously published in the USA and Canada
by Spon Press
29 West 35th Street, New York, NY 10001

Spon Press is an imprint of the Taylor & Francis Group

© 2002 Graham Merrington, Linton Winder, Robert Parkinson
and Mark Redman

Publisher's Note
This book has been prepared from camera-ready copy supplied
by the authors
Printed and bound in Great Britain by
TJ International Ltd, Padstow, Cornwall

All rights reserved. No part of this book may be reprinted or
reproduced or utilized in any form or by any electronic,
mechanical, or other means, now known or hereafter invented,
including photocopying and recording, or in any information
storage or retrieval system, without permission in writing
from the publishers.

British Library Cataloguing in Publication Data
A catalogue record for this book is available
from the British Library

363.7 288

Library of Congress Cataloging in Publication Data
A catalog record for this book has been requested

ISBN 0–415–27340–4 (hbk)

ISBN 0–419–21390–2 (pbk)

Contents

Figures

Tables

Acknowledgements

The authors would like to thank the following for the provision of material, helpful suggestions and comment: Tom Misselbrook, Institute of Grassland and Environmental Science, North Wyke Research Station. Helen Stokes, Matt Lobley and Mick Fuller at Seale Hayne, Department of Ecology, Lithuanian University of Agriculture, Noreikiskes. Environment Agency (North West) for data for case studies: Jeremy Joseph, John Quinton, National Soil Resources Institute, Cranfield University (Figure 4.1) Cam Grant and Brian Williams, Department of Soil Water, Adelaide University. Crown copyright material is reproduced under Class Licence Number C02P0000070 with the permission of the Controller of HMSO and the Queen's Printer of Scotland. Finally, the authors express personal thanks to their respective families for their encouragement and support throughout the project.

Agriculture and Pollution 1

1.1 SETTING THE SCENE

During the latter half of the twentieth century, the global human population doubled from less than 3000 million to 6000 million. As the global population increases, demand for food continues to rise. This leads to the intensification of agriculture which in turn places increasing demands on the natural environment (Brown *et al.*, 2000).

Agriculture is of fundamental importance to any national economy and the lifeblood of rural communities throughout the world. It occupies 35% of the world's land surface with 11% under direct cultivation and 24% managed as permanent pasture (UNEP, 1992). In the United Kingdom more than 76% of land is under agricultural production (MAFF, 2000a). The first evidence of agricultural activity in the UK can be traced back to 5000 BC (Reed, 1990), although our book is concerned with changes occurring only in the last 70 years or so. The UK agricultural industry has undergone a major revolution since the 1930s; progress in animal and crop breeding, the availability of pesticides and fertilisers, and ever-advancing technology has resulted in a substantial increase in productivity and levels of national self-sufficiency.

Successive government policies, notably the farm support measures of the Agriculture Act 1947 and adherence to European policy (Box 1.1) have provided UK farmers with markets for their products and a price structure that has encouraged agricultural intensification (MAFF, 1995, 2000a). This drive towards intensive production (Table 1.1) has led to a major increase in the use of agrochemicals like fertilisers (Figure 1.1) and pesticides. In parallel, there have been many technological advances adopted by the agriculture industry in recent years, such as the genetic modification of crop plants and precision farming.

Table 1.1 Changes in agricultural practice in the UK between 1930 and 1990 (Edwards and Withers, 1998).

	1930	1990
Total agricultural area ('000s of ha)	12,129	11,931
Arable land area ('000s of ha)	5,222	6,407
Autumn sown crops area (000s of ha)	778	2,896
Livestock density (head ha^{-1})	0.83	1.77

The area of land affected by agriculture, its reliance on natural processes and the use of technology to intensify production results in a unique system both economically and ecologically with the following four key characteristics (OECD, 1997):

- Economic viability of agriculture is influenced greatly by the natural environment. Productivity depends upon factors such as climate, soil fertility and water supply.
- Agricultural activities affect the quality of the environment. Crops and livestock form part of the agroecosystem, utilising natural resources for growth. Environmental benefits such as the maintenance of traditional landscapes may be apparent, but equally there may be costs such as deterioration in soil, water and air quality or the loss of habitats important for conservation.

- The relationship between agricultural activity and the environment is complex and site-specific. Interacting factors include the physical, chemical and biological attributes of the local environment, the mix of farm enterprises, management practices, and the production technologies adopted.

- Cultural and political influences affect the agriculture impacts on the environment. Most developed nations support food production by public subsidy and government intervention (OECD, 1996; MAFF, 2000a). Such measures inevitably affect the level of food production, its location and management.

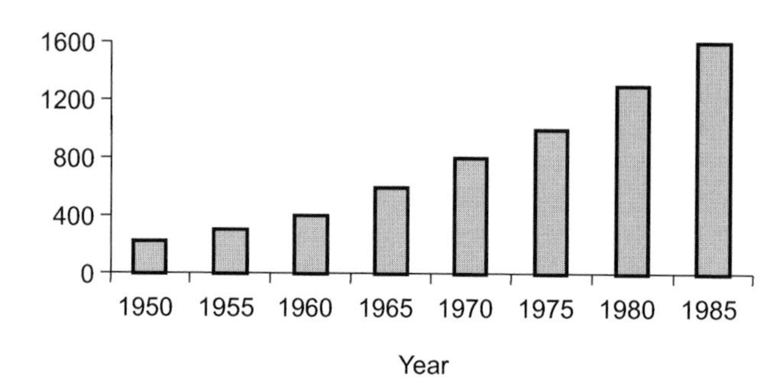

Figure 1.1 Annual use of nitrogen fertiliser in the UK ('000's of tonnes) (Winter, 1996).

Box 1.1 The European Union and agricultural intensification

Many of the changes in UK agriculture have been influenced by initiatives within the EU. The Common Agricultural Policy (CAP) was conceived by the original members of the European Community in order to encourage food production at a time when food shortage was still a recent memory. The original objectives contained in the 1957 Treaty of Rome included increasing productivity, stabilising markets, ensuring reasonable food prices for consumers, and maintaining a fair standard of living for farming communities. The plentiful supply of food and the extent of EU exports are a testament to its success. However, the CAP has been criticised because it created an economic climate in which food production was encouraged at the expense of other considerations such as the environment. Food surpluses were created that were dealt with either by being destroyed, or by export with subsidies (damaging international trade and affecting agriculture in developing countries).

The CAP has also reinforced the post-war trend in many European countries towards the expansion, intensification and specialisation of crop and livestock production. The price structure of markets reduced the economic risks associated with specialisation, stimulated the use of higher inputs to increase output, and encouraged the expansion of production into previously uncultivated areas. Farms grew larger, more capital-intensive and eager to adopt further improved technologies with which to intensify production (Clunies-Ross and Hildyard, 1992).

Escalating financial costs, environmental concerns, the approaching enlargement of the EU and the distorting effect of the CAP on world trade continue to move towards fundamental reform. The latest 'Agenda 2000' reforms will result in the internal market prices for cereal, beef and dairy products being reduced to the level of global market prices by 2006. In addition, the Rural Development Regulation (the so-called 'second pillar' of the CAP) provides further evidence of a long-term desire to shift support from production to rural development and environmental management (although close to 90% of the CAP budget is still devoted to commodity regime support and compensation payments).

This book focuses upon the UK, and is illustrated primarily with European examples. The need to identify, understand and solve the problems caused by agricultural activity is clear from statistics that show that it can be a major cause of pollution (Figure 1.2). Seven key types of agricultural pollution: nitrates, phosphates, sediment loss, organic wastes, gaseous emissions, pesticides and genetic modification are included. We describe why pollution may occur and how such problems can be overcome. Before we investigate these topics individually, the underlying causes of agricultural pollution and responses to them are considered. The final chapter reviews the role of agricultural policy, and reflects on how it may be used as a tool to deliver environmentally sensitive agricultural systems.

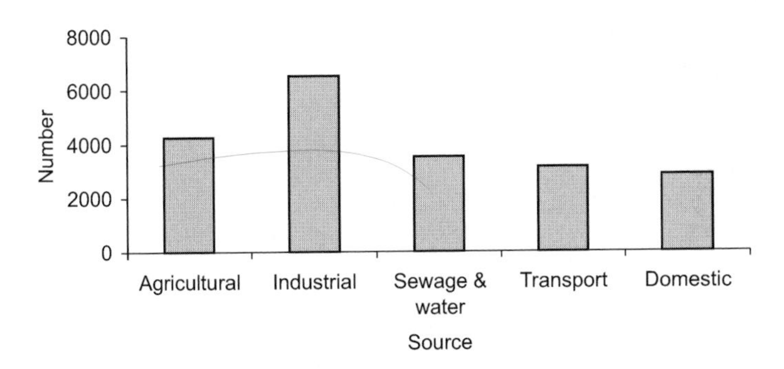

Figure 1.2 Water pollution incidents recorded within a range of economic sectors (Environment Agency, 2001).

Defining agricultural pollution

At its most inclusive, the term pollution can be used to describe all unwanted environmental effects of human activity. The Oxford English Dictionary defines pollution as 'the presence in the environment, or the introduction into it, of products of human activity which have harmful or objectionable effects'. This definition could include 'visual' pollution such as unsightly farm buildings (Conway and Pretty, 1991) but we use a definition whereby a pollutant is a substance, including those:

- deliberately introduced into the environment (e.g. pesticides, fertilisers, genetically modified crops and sewage sludge);
- produced by agricultural processes as wastes (e.g. silage effluent and livestock slurry);
- produced by the enhancement of natural processes in the course of agricultural activity (e.g. increased nitrous oxide emissions from cultivated soils or soil erosion).

A further distinction between a 'contaminant', which is, any substance introduced by human activity into the environment with no evidence of harm, and a 'pollutant', which is causing damage or harm may be made (Crathorne *et al.*, 1996). Substances entering the environment may only cause pollution if they are:

- present in excessive quantities – 'concentration effects';
- in the 'wrong place at the wrong time';
- transformed into harmful 'secondary pollutants' as a result of biological or chemical processes.

Pollutants from agricultural systems have the potential to have a major impact, ranging from the immediate on-farm environment to food products at the point of sale, and from local groundwater sources to the stratosphere. The impacts of agricultural pollution can be categorised into the following areas (OECD, 1997):

- quality of natural resources, notably the physical, biological and chemical condition of soil, water and air;
- composition and functioning of terrestrial, aquatic and marine ecosystems, including issues of biodiversity and habitat quality;
- other environmental impacts such as public nuisance caused by odours from livestock production.

The occurrence of pollution in agricultural systems is well documented (Baldock *et al.*, 1996; OECD, 1997; Edwards and Withers, 1998; Isherwood, 2000; EFMA, 2001) and the precise effect on the natural environment is mediated by the interaction between environmental factors and farm management. The move towards more intensive farming methods has led to a marked increase in the number of pollution incidents recorded (Figure 1.3) and this has driven the development of pollution control strategies designed to reverse this trend.

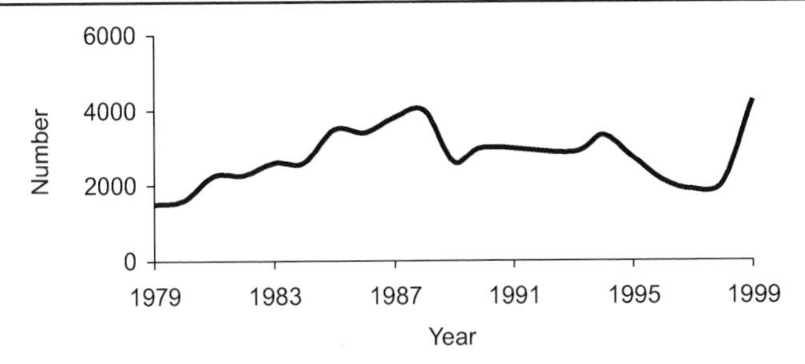

Figure 1.3 Number of agricultural pollution incidents recorded between 1979 and 1999 in England and Wales (NRA, 1992; Environment Agency, 2001).

This book addresses agricultural pollution although the full environmental and socio-economic impact of post-war agriculture also includes declines in farmland wildlife, the loss of traditional landscapes, rural depopulation, animal welfare, and health related issues (Baldock *et al.*, 1996; Corpet, 1996; DoE, 1996; Jones, 1999; MAFF, 2000a, 2001).

In this book we consider the following environmental issues:

- contamination of ground and surface water by nitrates (Chapter 2), phosphates (Chapter 3), organic wastes (Chapter 5) and pesticides (Chapter 7), all of which can disrupt aquatic and marine ecosystems and have significant effects on drinking water quality;
- disruption of agroecosystems by pesticides (Chapter 7) or genetically modified crops (Chapter 8), including flora and fauna in crops and semi-natural habitats;
- the effect of sediment loss due to erosion (Chapter 4);
- contamination of soil and crops by metals, organic micropollutants and pathogens from the application of livestock wastes and non-agricultural industrial wastes (Chapter 5);
- atmospheric contamination by ammonia, methane and nitrous oxide which play various roles in acid rain production, global warming and ozone depletion (Chapter 6).

1.2 SOLVING THE PROBLEM

Monitoring impacts

Assessment of the environmental consequences of agricultural pollution is needed before solutions to the problems can be formulated. Agriculture is only one of many economic activities that cause pollution and so its 'share of responsibility' for any environmental impact incurred must be identified before control priorities are set. For example, pesticide pollution is not just caused by agriculture. Most of the herbicides commonly detected in water also have non-agricultural uses with local authorities and public utilities regularly spraying public parks, roadside verges, railway lines and playing fields to control weeds. Until their prohibition from non-agricultural use in 1992, this included the use of atrazine and simazine, two of the most commonly detected herbicides in drinking water. Further investigation of the effects of agricultural pollution is complicated by:

- The complexity of contamination and pollution pathways: for example, pesticides may affect wildlife by direct contact with the chemical or its breakdown products, indirectly by contamination of food sources, or by the destruction of habitats and resources upon which species depend.

- Spatial and temporal variations in the occurrence of harm or damage: There are essentially two main sources of pollution from agriculture: 'point', involving discrete and easily identifiable incidents such as leakage from a slurry store or silage clamp, and 'diffuse', involving the leaching and run-off of pollutants from large areas of agricultural land to ground and surface water.

- The unpredictable nature of pollution due to variations caused by climate, soil type and other environmental factors.

The challenge is to identify the effects of pollution and then translate this understanding into appropriate action to manage the problem. Common responses (OECD, 1997) include:

- government action through changes in policy and law, research and development, training and information programmes and economic instruments such as financial subsidies and taxes;

- responses by the agricultural industry, such as the voluntary adoption of new quality standards and the imposition of stricter quality;

- modified behaviour by farmers, including changes in the use of agrochemical inputs and other farm management practices;

- consumer reactions expressed via changing patterns of purchase and consumption.

Pollution management can be achieved in two ways. Firstly, we can attempt to 'cure' the problem by acting against the pollutants themselves (e.g. by water treatment). Secondly, we may 'prevent' the problem by addressing the underlying causes of pollution (e.g. by encouraging the adoption of alternative agricultural practices that are less polluting). In both cases, the actions that can be taken are technical; whether they are adopted depends upon the presence of appropriate knowledge, effective legal regulations and adequate financial incentives (Conway and Pretty, 1991; MAFF, 1998a,b). This book includes solutions that are essentially farm-based, modifying management practices and business decisions via:

- the availability of new technologies;
- the provision of information and advice to encourage 'good agricultural practice';
- statutory controls and regulations that enforce change;
- agri-environmental policy and the provision of financial incentives in the form of taxes or subsidies;
- the emergence of alternative agricultural systems with a market linkage offering financial incentives in the form of price.

For each pollutant we describe current practical solutions to control their impact by considering new technology, good agricultural practice and regulation. We also describe the development and implementation of contemporary agri-environmental policy in the UK and western Europe, including the emergence of alternative agricultural systems such as Integrated Crop Management (Chapter 9).

Cost – the driver for pollution control

It has been recognised by government and industry that the 'external costs' or 'externalities' caused by economic activity should be considered when making decisions regarding pollution control (DoE, 1994). An externality is a side effect (or by-product) of agricultural practice which is unpriced within the economy of the farming system but which nonetheless incurs a cost for someone (or something) else by reducing their profit or welfare (Hanley, 1991). These externalities include:

- the depreciation of natural capital through the use of non-renewable natural resources such as oil and coal or the loss of other natural assets such as biodiversity and landscape;
- declines in personal or collective 'welfare' such as public health;
- the cost of environmental degradation including the cost of cleaning up damage;
- the cost of defensive expenditure including the cost of preventative action to avoid environmental damage.

The principal challenge when accounting for agricultural externalities is not in their identification, but the assignment of monetary value. Costs may be financial (e.g. incurred in water treatment) or economic (e.g. due to the loss of a landscape feature valued by people). Examples showing how costs may be evaluated are given in Box 1.2. It is a challenge for agricultural policy makers to facilitate the transition of farm practice to those which are environmentally sensitive, allowing farmers to modify their farming practices whilst maintaining the economic viability of their businesses.

Box 1.2 The costs of agricultural pollution

Nitrates in drinking water
The EC maximum admissible concentration (MAC) of nitrate in human drinking water is 50 mg l^{-1} (50 parts per million). However, an increasing number of raw water sources in the UK exceed this concentration and water supply companies have been forced to introduce treatment programmes (DWI, 2000). Treatment options (Croll and Hayes, 1988) include blending high nitrate content water with that which is less polluted, biological denitrification, ion exchange, reverse osmosis and electrodialysis. Estimates of the investment incurred in the installation of denitrification equipment ranges from £148 million to £200 million (DoE, 1986; Ofwat, 1992), with annual running costs of at least £10 million per year.

Phosphates in waters
The total polluting loads from sewage treatments works in the UK has fallen by between 30 and 40% during the 1990s and phosphate loads specifically by 37%. This has been due to improved sewage treatment with investment of £250 million combined with a reduction of phosphate usage in detergents (Environment Agency, 2000a,b). Agriculture is a diffuse source, accounting for over 50% of the Europe-wide phosphate in surface waters (Environment Agency, 2000b). Estimated annual costs for the removal of phosphate from surface waters from agriculture in the UK are in the order of £55 million (ENDS, 2000b).

Organic wastes
The production, storage and disposal of animal waste and silage effluent can present significant risks to the aquatic life of streams and rivers, although the total number of farm pollution incidents due to organic wastes is now declining in the UK. Nonetheless, the Environment Agency still spends approximately £5 million per year on surveying and correcting river pollution incidents caused by agriculture (National Audit Office, 1995).

Gaseous emissions
In the UK, agriculture is responsible for approximately 8% of all greenhouse gas emissions, in particular nitrous oxide and methane (MAFF, 2000a). Predicting the effects of increased global warming on climate change in the UK over the next 50 years is fraught with difficulty, especially with regard to changes in rainfall, storminess and extreme events such as drought (MAFF, 2000b). The possible costs of climate change upon agricultural production are also only predictions, with considerable uncertainty related to weed, pest and disease outbreaks, global commodity price effects and changes in yield and quality of arable crops. Nevertheless, the combined cost of the detrimental manifestations of nitrous oxide and methane emissions from agriculture have been estimated to be well over £1,000 million a year (ENDS, 2000a).

Pesticides in drinking water

Under the 1980 EC Directive on Drinking Water Quality, the MAC for any pesticide in drinking water, irrespective of its toxicity, is 0.1μg l^{-1} (0.1 part per billion). This is acknowledged as one of the most stringent pesticide standards in the world and is arguably very difficult to enforce. Since the early 1990s an increasing number of UK groundwaters providing sources of drinking water have pesticide concentrations in excess of the EC standard and it is currently estimated that 8% of all those waters tested exceed the limit set (MAFF, 2000a). This has triggered huge investment in treatment plants by water companies in an attempt to reduce pesticide levels in the water supplied to their customers (ENDS, 2000b). Removal of trace pesticides from water is complex and expensive; technology available includes granular activated carbon (GAC), ozone treatment and an activated carbon sandwich between layers of slow sand filters (ENDS, 2001). Estimates of the total capital investment undertaken by UK water companies on pesticide treatment plants range from £800 million to £1000 million (Ofwat, 1992). Annual running costs are also expected to have risen by about 10% of capital expenditure i.e. £80 million to £100 million per year.

The control of pollution can be viewed in the broader context of 'sustainable development'. This concept is founded upon the idea that the environment is a finite entity that is incapable of absorbing the impact of everything released into it or removed from it. In other words, the environment only has a certain 'capacity' to accommodate the impact of human activity. As people depend upon the environment for their survival and quality of life they have a duty of care to look after it for the benefit of themselves and future generations (Jacobs, 1991). Agriculture occupies a central place within the sustainability debate since it occupies more land than any other economic activity. To meet the challenge of sustainability, agriculture must reduce its environmental impact by minimising or eliminating pollution. In doing so it must also remain economically viable in order to survive and play its part in the life and economy of rural areas.

REFERENCES

Baldock, D., Bishop, K., Mitchell, K. and Phillips, A. (1996) *Growing Greener: Sustainable Agriculture in the UK*. Council for the Protection of Rural England and World Wide Fund for Nature, London.

Brown, L., Flavin, C. and French, H. (2000) *State of the World 2000*. Worldwatch Institute, 262pp.

Clunies-Ross, T. and Hildyard, N. (1992) *The Politics of Industrial Agriculture*. Earthscan Publications, London.

Conway, G. and Pretty, J. (1991) *Unwelcome Harvest: Agriculture and Pollution*. Earthscan Publications, London.

Corpet, D.E. (1996) Microbiological hazards for humans of antimicrobial growth promoter use in animal production. *Revue de Médecine Vétérinaire* **147**, 851–862.

Crathorne, B., Dobbs, A.J. and Rees, Y. (1996) Chemical Pollution of the Aquatic Environment by Priority Pollutants and its Control. In: *Pollution, Causes, Effects and Control* (Ed. R.M. Harrison). 3rd Edition. The Royal Society of Chemistry, Cambridge, pp. 1–25.

Croll, B. and Hayes, C. (1988) Nitrate and water supplies in the United Kingdom. *Environmental Pollution* **50**, 163–187.

DoE (1986) *Nitrate in Water: a Report by the Nitrate Co-ordination Group*. Department of Environment Pollution Paper No. **26**, HMSO, London.

DoE (1994) *Sustainable Development: the UK Strategy*. Department of Environment Command Paper **2426**, HMSO, London.

DoE (1996) *UK Indicators of Sustainable Development*. HMSO, London.

DWI (2000) *Overview of Water Quality in England and Wales, Drinking Water 1999*. Drinking Water Inspectorate, Department of the Environment, Transport and the Regions, London.

Edwards, A.C. and Withers, P.J.A. (1998) Soil phosphorus management and water quality: a UK perspective. *Soil Use and Management* **14**, 124–130.

EFMA (2001) *Sustainable Soil Management: an Achievable Goal*. European Fertilizer Manufacturers Association, Brussels.

ENDS (2000a) *The diffuse pollution challenge*. ENDS Report **310** (November).

ENDS (2000b) *Farming's environmental costs top £1.5 billion per year says Agency*. ENDS Report **309** (October).

ENDS (2001) *Water firms urged to end chlorine addition*. ENDS Report **314** (March).

Environment Agency (2000a) *Achieving the Quality, the Environment Agency's Views of the Benefits to the Environment of Water Company Investment over the Next Five Years*. Environment Agency, Department of the Environment, Transport and the Regions, London.

Environment Agency (2000b) *Aquatic Eutrophication in England and Wales: a Management Strategy*. Environment Agency, Department of the Environment, Transport and the Regions.

Environment Agency (2001) *Water Pollution Incidents 1999*. URL: http://www.environment-agency.gov.uk.

Hanley, N. (Ed.) (1991) *Farming and the Countryside: An Economic Analysis of External Costs and Benefits*. CAB International, Wallingford.

Isherwood, K.F. (2000) *Fertilizer Use and the Environment*. (Revised Edition). International Fertilizer Industry Association, Paris.

Jacobs, M. (1991) *The Green Economy*. Pluto Press, London.

Jones, D.L. (1999) *Escherichia coli* O157 in the environment. *Soil Use and Management* **15**, 76–83.

MAFF (1995) *European Agriculture: the Case for Radical Reform*. Ministry of Agriculture, Fisheries and Food, London.

MAFF (1998a) *Code of Good Agricultural Practice for the Protection of Water*. Revised 1998. Ministry of Agriculture, Fisheries and Food, London.

MAFF (1998b) *Guidelines for Farmers in Nitrate Vulnerable Zones*. Ministry of Agriculture, Fisheries and Food, London.

MAFF (2000a) *Towards Sustainable Agriculture: Pilot set of Indicators*. Ministry of Agriculture, Fisheries and Food, London.

MAFF (2000b) *Climate Change and Agriculture in the United Kingdom*. Ministry of Agriculture, Fisheries and Foods, London.

MAFF (2001) *Foot and Mouth Disease: Public Information Factsheet 1*. Ministry of Agriculture, Fisheries and Food, London.

National Audit Office (1995) *National Rivers Authority: River Pollution From Farms in England*. House of Commons Paper **235**. HMSO, London.

NRA (1992) *The Influence of Agriculture on the Quality of Natural Waters in England and Wales*. National Rivers Authority, Bristol.

OECD (1996) *Agricultural Policies, Markets and Trade in OECD Countries – Monitoring and Evaluation 1996*. Organisation for Economic Co-operation and Development, Paris.

OECD (1997) *Environmental Indicators for Agriculture*. Organisation for Economic Co-operation and Development, Paris.

Ofwat (1992) *The Cost of Quality – a Strategic Assessment of the Prospects for Future Water Bills*. Ofwat, Birmingham.

Reed, M. (1990) *The Landscape of Britain: from the Beginnings to 1914*. Routledge, London.

UNEP (1992) *The World Environment 1972–1992: Two Decades of Challenge*. Chapman and Hall (on behalf of the United Nations Environment Programme), London.

Winter, M. (1996) *Rural Politics: Policies for Agriculture, Forestry and the Environment*. Routledge, London.

Nitrates and Nitrogen Loss 2

2.1 INTRODUCTION

The present structure and output of agricultural systems could not be maintained without the advent and widespread use of synthetic or mineral fertilisers. Of the major plant nutrients N, not only provides the greatest responses in crop yield from fertiliser addition but is also the most readily lost from the agroecosystem.

In parallel with increased agricultural production over the last 50 years has been the increase in nitrate (NO_3^-) concentrations in rivers, lakes and underground aquifers. There is strong evidence to suggest that this is due to pollution from agriculture. Furthermore, this loss of N from agriculture as leachate in the form of NO_3^-, but also to a lesser extent as gaseous forms of N or erosion as N associations with soil particles, represents an economic shortfall, in that the applied N is not being utilised for food production. This chapter tackles three major questions:

- Why have NO_3^- levels increased in waters?
- Is this increase harmful to the environment?
- What can be done to reduce and stop further increases in NO_3^- pollution and losses from agricultural systems?

Nitrogen and agriculture

Nitrogen is an essential constituent of all nucleic acids, amino acids and proteins, and therefore fundamental to the reproduction and growth of all organisms. In a general introduction to the global N cycle, Jenkinson (1990) estimated that in 1990 the world's human population contained a total 10 million tonnes of N. Although this is small compared to the total amounts of N in the atmosphere (3.9×10^9 million tonnes), soil organic matter (1.5×10^5 million tonnes) or plants (1.5×10^4 million tonnes), the human population is increasing and is inextricably linked via agriculture and the food industry to the global N cycle (Schlesinger, 1997).

Nitrogen occurs naturally in soils and is closely associated with soil organic matter. However, it is the simple ionic forms of ammonia (NH_4^+) and NO_3^- which plants can easily absorb and utilise (Whitehead, 2000). Of the elements essential for plant growth, N is required in the greatest quantity by agricultural crops (Table 2.1). The exceptions to this rule-of-thumb are those crops which form large underground storage organs, notably potatoes, since these also require large quantities of phosphorus (P) and potassium (K).

Nitrogen-containing compounds are involved in virtually all of the biochemistry of the crop plant. This includes chlorophyll that is essential for photosynthesis, the nucleic acids in which the pattern for the plant's growth and development is encoded, and a variety of plant proteins ranging from lipoprotein membranes to enzymes such as ribulose 1,5-biphosphate carboxylase-oxygenase (Rubisco) which plays a key role in the conversion of atmospheric carbon dioxide (CO_2) into organic carbon during photosynthesis.

Nitrogen deficiency is generally characterised by the yellowing or chlorosis (i.e. loss of chlorophyll) of the lower leaves of crop plants, starting from the tip and extending to the whole leaf with increasing deficiency. In severe cases, the whole plant is stunted and the leaves remain small.

Table 2.1 Typical nutrient removal (kg ha^{-1}) in a harvested winter wheat crop nutrient offtake per tonne of plant material.

	Yield (tha^{-1})	N	P	K	S	Mg	Ca
Winter wheat	8	168	29	24	13	10	3

While the timely application of relatively small amounts of additional NH_4^+ and NO_3^- to the soil will often relieve the symptoms of sickly-looking crops, much higher levels of mineral N are required to significantly increase crop yield. An abundant supply of mineral N increases crop yield by influencing leaf area in two distinct ways:

- By encouraging the rapid growth of above-ground vegetation: This can have an important influence upon crop yield, since any increase in the size of the crop canopy promotes both the interception of sunlight and the absorption of CO_2 thereby increasing photosynthetic efficiency.

- By promoting the duration of the crop canopy: Nitrogen is very mobile within the plant and any shortage of N in young tissue is usually met by the mobilisation of N from the older leaves resulting in their chlorosis and eventual death. An abundant N supply avoids this problem and maintains more leaves to carry on photosynthesising for longer.

In certain crops, a good supply of mineral N may also help to improve the quality, as well as the yield, of the final harvested produce. For example, bread-making wheat varieties need to contain at least 11–12% protein in order to form a satisfactory dough. This requires a sufficiently high N supply (late in the growing season) to achieve an N concentration in the grain of around 2.0% on a dry matter (DM) basis. In other crops, increased supplies of mineral N are not so welcome. The availability of too much N for potato plants can produce too many over-large tubers, while barley grown for malting needs grain with as much starch and as little protein content as possible (ideally not in excess of 1.6% N). Too much N may also reduce the sugar content of sugar beet (Isherwood, 2000). Excessive amount of N also produces vegetative growth with large succulent thin-walled cells. This can cause two problems:

- the leaves and stems are more readily attacked by insect pests and fungal diseases;

- the stems are less mechanically strong and crops are subsequently prone to 'lodging' i.e. being blown over in wet and windy weather.

Despite being the most important crop nutrient in agriculture, N is also an environmental pollutant causing significant ecological disturbance. This chapter will consider the 'leakage' of N from agriculture into the wider environment in detail and discuss those factors influencing the occurrence and behaviour of N within agricultural systems.

2.2 THE NITROGEN CYCLE

Nitrogen is a transient nutrient and the amount available in the soil at any one time to meet the demands of a growing crop is the product of the complex network of physical, biological and chemical pathways through which the various forms of N move:

- into the soil (inputs);
- within the soil (transformations); and
- out of the soil (losses).

Together these pathways are known as the soil N cycle (Figure 2.1) and are an integral part of the overall cycling of N within nature. A full discussion of this cycle and its significance for agricultural production is beyond the scope of this book, but it has been extensively reviewed elsewhere (e.g. Wild, 1988; Powlson, 1993; Tisdale *et al.,* 1993).

Nitrogen can enter the soil N cycle in a number of ways and in different forms. Atmospheric deposition, biological N_2 fixation, fertilisers and animal feeds effectively import N from outside of the farm, whilst animal manures typically transfer N from one part of the farm (e.g. where livestock which have been overwintering in sheds and barns) to another.

Atmospheric deposition

Agricultural land receives significant quantities of N via the deposition of N oxides and NH_3 from the atmosphere. The major source of N oxides in the atmosphere is fuel combustion and the so-called NO_x emissions from power stations and motor vehicles. Atmospheric NH_3 is derived from a number of sources including industrial emissions, coal burning, livestock wastes and other agricultural sources (Schlesinger, 1997) (Chapter 6).

Since most atmospheric N compounds are highly soluble in water, deposition may occur in rainfall, although the dry deposition of gaseous and particulate material may also occur. An estimated 30–50 kg N ha^{-1} $year^{-1}$ are deposited on agricultural land from the atmosphere in southern and eastern England (and the same for some parts of Germany). Of this, when deposited onto land in cereals, it is estimated that 5% is leached, 12% denitrified, 30% immobilised and the remaining taken off by the crop (Goulding *et al.,* 1998). Total N deposition in intensive livestock production area of the Netherlands may range from 40–80 kg N ha^{-1} $year^{-1}$ (Whitehead, 2000). For the farmer this may represent a useful N input, but may be particularly detrimental in natural and semi-natural ecosystems (Box 2.1). Even in areas remote from intensive production agriculture deposition may still be in the region of 15 kg N ha^{-1} $year^{-1}$ (Brady and Weil, 1999).

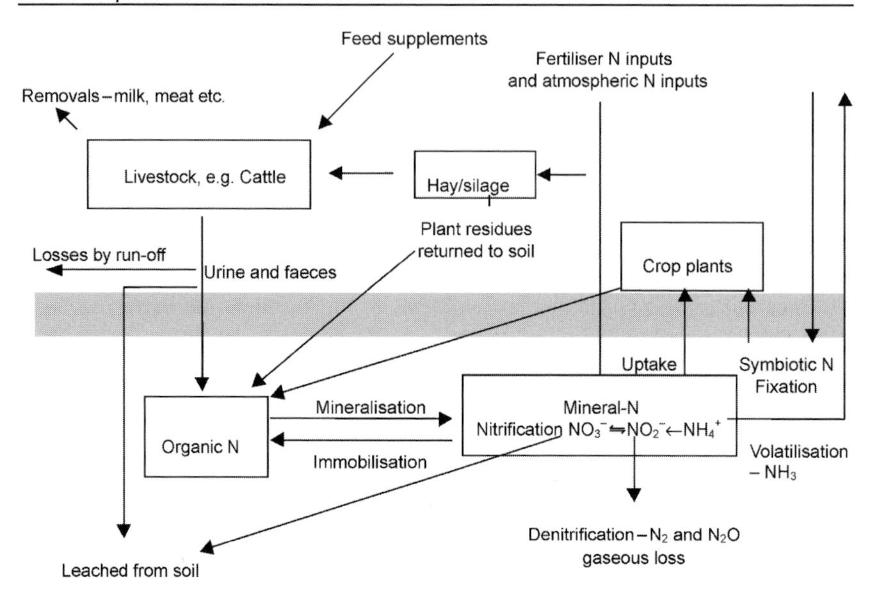

Figure 2.1 The agricultural N cycle (adapted from Rowell, 1994).

Biological fixation

Certain species of bacteria and algae are capable of reducing atmospheric N to NH_3. The most important agricultural example is the *Rhizobia* bacteria that form a close symbiotic relationship with leguminous crops such as peas, beans and clover. Legumes are mainly grown on the farm as either arable or forage crops. Arable legumes (e.g. field peas and beans) are grown and harvested to produce a dried grain for inclusion as a protein source in the diet of farm animals, while forage legumes (e.g. red and white clover) are commonly grown mixed with grass, and are either used for livestock grazing or cut for hay and silage. Legume crops usually leave large quantities of high protein content crop residues which can make a significant contribution to levels of organic N in the soil and, upon decomposition and mineralisation, to soil mineral N levels and the growth and yield of subsequent crops. Some legume crops are grown specifically for incorporation into the soil as green manures although in recent times their use in conventional agriculture has diminished (Parsons, 1984).

In the UK crop rotations including leguminous crops formed the traditional base of agriculture for many centuries. One of the best known rotations in the 1700s was the 'Norfolk four course'. This originally took the form of roots, barley, seed and wheat. The seed component of the course of the rotation was some form of legume, notably a one-year red clover ley (sometimes with ryegrass) or an arable legume crop. In some parts of the country, the 1-year seeds crop was extended into a short-term 2-year ley, or into the medium to long-term leys which form the basis of traditional ley/arable farming systems e.g. 4–5 year forage legume or grass/clover ley followed by up to 3 years cereals (Laity, 1948).

Biological N_2 fixation by legumes is highly variable depending upon the number of active nodules, their size and longevity, and the bacterial strains occupying them. These factors in turn are affected by the complex interaction of legume species and cultivar, crop management and conditions of growth (notably water availability and soil nutrient status). In an extensive review of North American work, LaRue and Patterson (1981) quoted estimates of annual N_2 fixation in the range of 10–100 kg N ha^{-1} for arable legumes and 100–250 kg N ha^{-1} for forage legumes. In Europe, productive grass-clover swards would be expected to fix between 100 and 300 kg N ha^{-1}, without the addition of N fertiliser (Whitehead, 2000).

Fertilisers

The industrial fixation of atmospheric di-nitrogen gas (N_2) is directly analogous to biological fixation since it also involves the reduction of N_2 to NH_3. During the commercial manufacture of NH_3 (Haber-Bosch process), hydrogen and atmospheric N_2 are combined at high temperature (300–500 °C) and pressure (400–1000 atmospheres) in the presence of a catalyst. The NH_3 produced may be used as a fertiliser material itself (Table 2.2), but is more commonly processed to fertiliser materials such as ammonium nitrate or urea (so-called 'straight' N fertilisers), or mono- and di-ammonium phosphates used for the manufacture of 'compound' NPK fertilisers.

Compared to the use of legumes as an N source, fertilisers directly supplement soil mineral N levels with NH_4^+ and/or NO_3^-, and therefore rapidly increase the amount of N available for crop uptake. As a guide (and assuming that the fertiliser is applied at an appropriate time), the yield of a wheat crop may increase by a maximum of 24 kg for each additional kilogram of fertiliser N applied, up to where the response begins to plateau (Figure 2.2) (Mackenzie and Traureau, 1997). As application rates are increased, a point of 'optimum' application is reached at which the availability of extra N ceases to be worthwhile for crop growth and is thus surplus to requirement (Figure 2.2).

Table 2.2 N-containing chemicals and materials commonly used in 'straight' and 'compound' fertilisers (White, 1997).

		N content (%)
'Straight' fertilisers	Ammonium sulphate	21
	Ammonium nitrate	35
	Calcium nitrate	17
	Urea	46
	Anhydrous ammonia	82
	Aqueous ammonia	25–29
'Compound' fertilisers	Mono-ammonium phosphate	11–12
	Di-ammonium phosphate	18–21

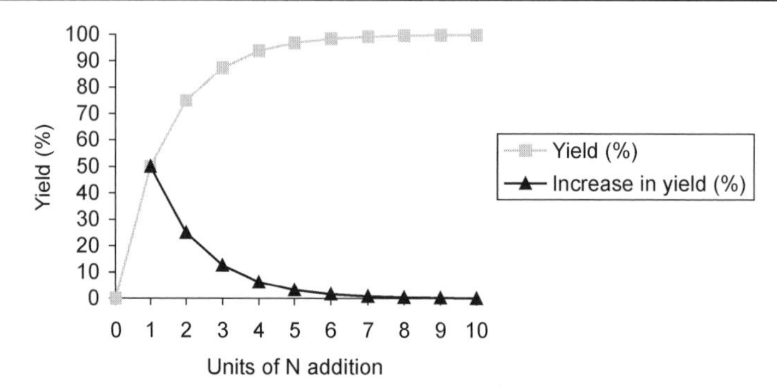

Figure 2.2 Relative yield from the addition of a growth factor, such as N, as described by the Mitscherlich Equation (data from Tisdale *et al.,* 1993).

Animal manures

Of the N consumed by livestock in the form of herbage and concentrate feeds, a relatively small proportion is actually utilised for the production of meat or milk (Whitehead *et al.*, 1986). Instead, the majority (typically 70% of the N consumed by cattle and 80% consumed by sheep) of N is excreted as dung and urine (Haygarth *et al.*, 1998). When cattle and sheep are grazing this excreta is voided directly to the soil surface, but once the animals are housed in the winter it accumulates in yards and buildings as slurry and manure which needs to be stored and spread on the land at an appropriate time (Chapter 5).

The amount of N available to crops following the surface application of manure varies with a range of factors, including soil type, form and source of manure and time of year. Figure 2.3 shows the percentage of total N available to the next crop following the winter and autumnal application of different manure types to a sandy soil. The usefulness of this type of information to farmers attempting to account for the fertiliser value of manures and reduce N leakage will become evident as the chapter continues (Box 2.2).

Manure N comprises two major fractions of agronomic interest, readily available inorganic N (mainly NH_4^+) and organic N (Chapter 5) (MAFF, 2001). Ammoniacal N is water soluble, comprising of urea and NH_4^+, and when the manure/slurry is applied to the soil supplements soil mineral N levels in a similar manner to fertiliser N. The organic N fraction must undergo mineralisation before being available for crop uptake.

2.3 NITROGEN TRANSFORMATIONS

Within the soil N cycle, a number of important biological and chemical transformations occur (Figure 2.1) which influence both the amount of N available for crop uptake and that which is at risk of 'leaking' into the wider environment. Two of the most important of these are mineralisation and immobilisation that involve the transformation of N between organic and inorganic forms in the soil.

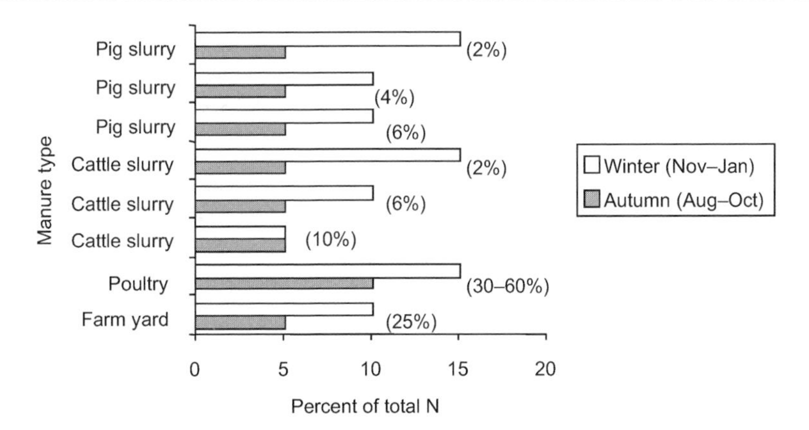

Figure 2.3 The percentage of total N available to a crop following surface application of manure. Numbers in parentheses are percentage DM content of the respective manure (data from Chambers *et al.*, 2000).

Nitrogen mineralisation

Soils naturally contain N in two discrete pools: organic and inorganic. Depending upon the cropping history of the soil, mineral composition and prevailing environmental conditions the sum of these two pools would be between 2000 and 6000 kg N ha^{-1}. Almost all of which would be in the organic form and therefore unavailable for crop uptake (Powlson, 1993). Soil organic matter is colonised by a variety of heterotrophic soil organisms that derive their energy for growth from the decomposition of organic molecules. During decomposition essential nutrient elements, including N, are also converted from organic to inorganic forms. This is termed mineralisation and occurs whenever soils are moist and warm enough for microbial activity, with a 'flush' of intense activity usually occurring in the spring and to a lesser extent in the autumn.

The mineralisation of organic N involves the degradation of proteins, amino acids, nucleic acids and other nitrogenous compounds to NH_4^+. Once formed, NH_4^+ joins the mineral N pool (along with NH_4^+ derived directly from fertiliser, manure and atmospheric deposition) and is subject to a number of potential fates, including further transformation by immobilisation, nitrification or adsorption/fixation, as well as direct loss from the soil by crop uptake or volatilisation. Mineralisation is the key process controlling N availability for plants and leaching loss (Goulding, 2000).

Nitrogen immobilisation

Nitrogen immobilisation is defined as the transformation of inorganic N compounds (NH_4^+, NO_3^- and NO_2^-) into organic N forms (Jansson and Persson, 1982). It occurs when there is a readily available source of carbon (C)-rich material, such as following the addition of crop residues like cereal straw which are low in N, but high in C (C/N ratio > 80:1, Chapter 5). Both mineralisation and immobilisation are greatly influenced by the availability of C (Powlson, 2000). As micro-organisms attack and decompose the C-rich material, they also absorb NH_4^+ from the soil and rapidly convert it to microbial biomass. Subsequently, the microbial biomass dies, enters the active phase of the soil organic matter and becomes liable to decomposition again. Some of the microbial N may enter the passive phase to form humus, or it may be released as mineral N. The continuous turnover of NH_4^+ in this manner forms a sub-cycle within the overall N cycle of the agricultural soil and produces a net effect – net mineralisation or net immobilisation – which influences the supply of NH_4^+ for other N cycle processes.

Nitrification

When (as is usual in most soils) the microbial population is limited by available C, most of the mineralised NH_4^+ is oxidised rapidly to NO_3^- by the process of nitrification. This is a two-stage process mediated by two important groups of bacteria: *Nitrosomonas* that oxidise NH_4^+ to NO_2^- and *Nitrobacter* which oxidise NO_2^- to NO_3^-. Since the oxidation of NO_2^- is more rapid than that of NH_4^+, there are only ever trace amounts of NO_2^- in the soil. Importantly, the mineralisation of organic N in soil and in crop residues is seen as one of the major sources of NO_3^- in agriculture (Powlson, 2000).

Adsorption and fixation

Ammonium can be adsorbed onto the surface of clay minerals and soil organic matter (the 'cation exchange sites'), from where it is freely exchangeable with other cations in the soil solution. Ammonium is also approximately the same size as the K^+ potassium ion and readily enters the interlayer portions of clay minerals (e.g. vermiculite). The collapse of this interlayer space, for example by drying, effectively fixes the NH_4^+, making it only very slowly available to enter the soil solution. This is thought of as being a partial reversal of the weathering processes undertaken by alumino-silicate minerals.

Nitrogen uptake and losses

There are a number of pathways by which N is lost from an agricultural soil. The most desirable route is via crop uptake and subsequent removal by grazing or harvest since this produces both an economic return for the farmer, and is not a direct cause of pollution. The proportion of applied N taken during a growing season should, under controlled conditions, be between 50 and 70% (for phosphorus < 15% and potassium 50–60%) (Isherwood, 2000). However, in practice, any N which is available in the soil for crop uptake at any time is also vulnerable to loss from the soil (without economic return and as a potential pollutant) in any one of three ways: NO_3^- leaching; gaseous emissions during denitrification and nitrification, and NH_4^+ volatilisation (Chapter 7). The importance of each of these loss pathways vary considerably and depends on the agricultural system.

Nitrate leaching

Due to the chemical and biological processes described previously, NO_3^- is generally the most prevalent form of soil mineral N in well-drained and aerated soils. Much of this NO_3^- is derived from the nitrification of NH_4^+, but levels are also supplemented directly by applications of fertiliser and, to a lesser extent, atmospheric deposition. When crops are growing rapidly they take up NO_3^- very quickly, thus reducing its susceptibility to loss. However, once plants stop growing any NO_3^- that continues to be made available in the soil (e.g. by mineralisation or the application of fertiliser) is vulnerable to losses, such as leaching.

Nitrate is a very soluble anion and, unlike NH_4^+, is not readily adsorbed or fixed by the soil. It may be considered as behaving 'conservatively' in the soil, in that it does not readily react with the soil mineral and organic surfaces and so tends to remain in soil solution in relatively high concentrations (cf. phosphate, Chapter 3). Therefore, NO_3^- moves through the soil predominantly by the mechanism of mass flow, which is the movement of ions in soil solution down hydraulic gradients, which may be created by a transpiring plant (Tisdale *et al.*, 1993).

Agricultural soils can be prone to 'leak' substantial quantities of NO_3^- in drainage water and this is frequently cited as one the main causes of N loss from soils (>40 kg N ha^{-1} in many UK soils) (Lord and Anthony, 2000). Major leaching losses occur when two conditions are met:

- Soil water movement is large i.e. the influx of water (either rainfall or irrigation) is greater than the evapotranspiration. This is further influenced by soil texture and structure that may affect the hydraulic conductivity and water storage capacity of the soil. Nitrate leaching losses are generally greater from poorly structured sandy soils than well structured clay soils.

- Soil NO_3^- levels are high due to the mineralisation of organic N (possibly from the application of animal manures), or the presence of excessive or unused fertiliser. For example, a September application of animal manure at the maximum recommended rate of 250 kg (total) N ha^{-1} (MAFF, 1998a) on a sandy soil would supply just 14 kg N ha^{-1} to a winter cereal crop, but 9 kg N ha^{-1} may be lost by leaching (MAFF, 2001). The proportion of leached NO_3^- derived from organic or fertiliser N sources will depend upon environmental conditions, the rate of applied N and the crop management systems employed. However, considerable quantities of NO_3^- leached from arable soils can originate from the rapid mineralisation of organic N rather than directly from applied fertiliser. Particularly under horticultural crops, where large N residues may be left in the soil (>300 kg N ha^{-1} may remain after some brassicas) (Rahn *et al.,* 1996). Figure 2.4 gives an indication of the variation in NO_3^- levels in a soil under arable cropping conditions in the UK and illustrates the time leaching and leakage is most likely to happen (Davies, 2000). A comparison of this Figure with Figure 2.3 highlights the importance of time of year in regard NO_3^- leaching from manure or fertiliser applications.

Seasonal rainfall and evapotranspiration patterns interact with soil NO_3^- levels to affect leaching losses. Although these vary greatly from year to year, and between regions, some general statements can be made based upon research in the UK and Denmark (Powlson, 1988; Addiscott, 1996; Simmelsgaard, 1999; Chambers *et al.*, 2000):

- in summer evapotranspiration generally exceeds rainfall and leaching is usually minimal. However, fertiliser N losses can occur if application coincides with intense, heavy rainfall;
- a particularly dry summer can limit crop N uptake and lead to the accumulation of NO_3^- in the soil which is then susceptible to autumn/winter leaching;
- provided that the soil is approaching or has returned to field capacity, autumn rainfall will leach any NO_3^- remaining from pre-harvest fertiliser application or derived from late summer/autumn mineralisation. Applications of fertiliser N to the seedbed of autumn sown crops are also very susceptible to leaching;
- in winter there is a large excess of rainfall over evapotranspiration and any NO_3^- present in the soil profile is readily leached;
- spring applied fertiliser N is susceptible to leaching if application coincides with heavy rainfall or NO_3^- is not rapidly removed from the available pool by crop uptake.

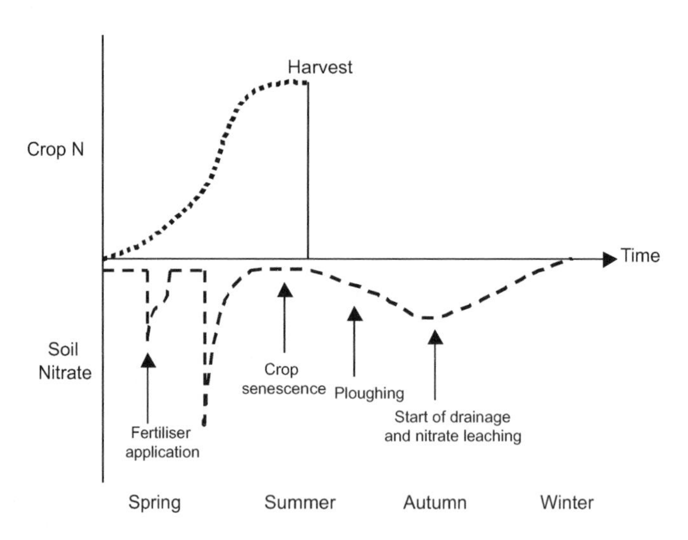

Figure 2.4 Variation in soil NO_3^- levels under a cereal crop over a period of one year (Davies, 2000).

The leaching of soluble organic nitrogen forms has also been suggested as being a major pathway of N loss from agricultural soils. It is thought that in soils this pool may be of similar size to the mineral N pool, but it is less subject to fluctuation and change (Murphy *et al.*, 2000).

Nitrous oxide and nitric oxide emissions from nitrification and denitrification

Another route by which significant amounts of N may leak from an agricultural soil is via gaseous emissions of nitric oxide, NO, nitrous oxide, N_2O, and molecular nitrogen, N_2 (Fowler *et al.*, 1996). The predominant source of these emissions is microbial nitrification (see above) and denitrification.

Denitrification is the dissimilatory reduction of NO_3^- is the major natural process by which oxidised N is returned to the atmosphere (Chapter 6) (Royal Society, 1983). It occurs under anaerobic soil conditions, when NO_3^- replaces O_2 as the terminal electron acceptor in microbial respiration. Conditions favouring denitrification are the presence of: adequate NO_3^- levels, denitrifying organisms, high soil water contents (low soil oxygen) and poor soil structure (both of which result in low air filled porosity and potential anaerobic conditions) (De Klein *et al.*, 2001). The main rate determining factors are soil temperature and the amount of readily available carbon substrate present (Smith and Arah, 1990) (Chapter 6).

Denitrification can be a very important N loss process in agricultural soils, particularly from heavy soils in wet conditions, but its measurement is frequently complicated by high spatial and temporal variability and difficulties of determining total denitrification loss from the measurement of N_2O emissions (Folorunso and Rolston, 1984; Arah *et al.*, 1991; De Klein *et al.*, 2001).

As well as representing a further economic loss to the farmer by reducing the availability of mineral N for crop uptake, emissions of NO and N_2O are pollutants that may pose an environmental hazard (Chapter 6) and also cause damage to natural and semi-natural ecosystems (Box 2.1).

Box 2.1 Critical nitrogen load

Critical load is the maximum atmospheric 'pollutant' load that sensitive ecological systems can tolerate without incurring long-term harmful effects (RCEP, 1996). The concept has mainly been applied to acid pollutants, indicating the capacity of soils in sensitive environmental areas to buffer atmospheric acid inputs.

Critical loads for N inputs are less easy to determine (N compounds, such as NH_3, can both acidfy and act as a nutrient), but have been estimated at 5–45 kg ha^{-1} $year^{-1}$ for a range of ecosystems from heathland to commercial forestry; the poorer the soil and the sparser the vegetation, the smaller the critical load.

An annual deposition of up to 40 kg N ha^{-1}, as estimated by Goulding *et al.* (1998), suggests that some natural and semi-natural ecosystems are likely to be receiving much more N than their critical load (Asman *et al.*, 1998). This will cause changes in the flora and fauna of the ecosystems due to the increased N supply, and increased soil acidification as the NH_3 deposited is nitrified and releases H^+ ions into the soil solution. Localised N deposition and soil acidification has, for example, been noted in field crops due to NH_3 volatilisation from a nearby intensive poultry unit (Speirs and Frost, 1987).

Ammonia volatilisation

Although NH_4^+ is not generally at risk of leaching because of its retention in the soil on negatively charged cation ion sites, gaseous losses of NH_3 do occur in agricultural systems, including emissions from soil (Figure 2.1). In Europe, the largest source of atmospheric pollution by NH_3 is agriculture, although oceans and biomass burning are also important and an estimated 60% of global emissions are from anthropogenic sources (Fowler *et al.,* 1996; Asman *et al.,* 1998).

Gaseous losses of NH_3 are most significant from agricultural systems involving livestock (especially intensive production systems) due to the breakdown of urea in animal urine and faeces (Whitehead *et al.,* 1986). Indeed, 92% of all ammonia in Western Europe originates from agriculture of which about 30% is from livestock (Isherwood, 2000). Ammonia loss is likely to be greatest where high concentrations of ammoniacal N occur. For example, from urine or slurry. Losses of N as NH_3 may be as great as 80% of the total when slurry is surface applied to grassland (depending on weather and sward conditions) (Whitehead, 2000).

In arable soils, the greatest losses occur when ammoniacal fertilisers or urea are applied under alkaline conditions. Ammonia losses resulting from surface volatilisation are aggravated by high soil temperatures and drying conditions, but can largely be prevented by placing fertilisers below the soil surface or working them in thoroughly with the top soil (Tisdale *et al.,* 1993).

2.4 NITROGEN FERTILISER USE IN AGRICULTURE

Sustained agricultural production depends upon the continual fixation of atmospheric N_2 to replenish the N lost from the soil in harvested crops, livestock production and the natural soil loss processes, such as leaching and denitrification, already described above.

Since the late 1940s agriculture has undergone rapid modernisation as political and economic support created a favourable economic climate in which technical efficiency and technological advancement were encouraged and flourished. In the modern agricultural systems now typical of areas such as northern Europe, reliance upon legumes has long been superseded by the use of industrial N_2 fixation and the application of synthetic or mineral N fertilisers. For example, in 1850, wheat yields in France where 1000 kg ha^{-1}, by 1950 this had reached 1600 kg ha^{-1}, with a fertiliser input of 1.1 million tonnes. By 1994–1996 yields had reached 6772 kg ha^{-1} with an input of total fertiliser of 4.8 million tonnes (of which 2.4 million tonnes was N) (Isherwood, 2000). A similar pattern has occurred in the UK (Figure 2.5), but importantly since 1982 N fertiliser application to winter wheat have remained relatively constant (180 kg N ha^{-1}), whereas as yields have increased significantly (from <6.0 to >7.5 t ha^{-1}) (FMA, 2000). This is thought to be largely due to technological improvements in other aspects of crop production including advances in seed strategies.

According to Wild (1993), three developments were important in establishing the place of fertilisers in modern agriculture:

- Long-term field experiments (e.g. those at the Rothamsted Experimental Station) showed that crop yields could be maintained with continuous cropping when the required plant nutrients were applied as fertiliser. In practice, good management also requires at least some crop rotation (e.g. inclusion of break crops for pest and disease control), as well as the addition of organic matter in some form.

- The industrial synthesis of N fertilisers in the 1920s, based upon the 'Haber process', made large-scale fertiliser production possible. Later developments in the oil industry provided cheap energy for the process.

- The introduction of higher-yielding, mainly short-stemmed, varieties of cereal crops which has made it profitable to add greater amounts of fertiliser. Improved cultivations, irrigation and the use of pesticides have also justified the use of more fertiliser.

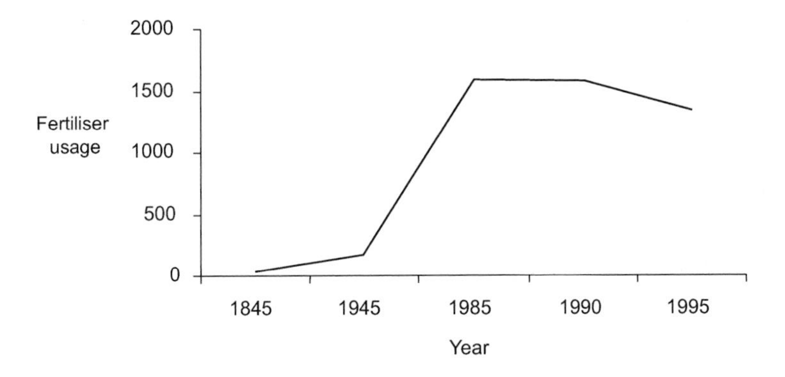

Figure 2.5 Nitrogen fertiliser use in the UK ('000s tonnes) (FMA, 1998).

Global fertiliser N use has dramatically increased since the mid-1940s and by the late 1980s, average fertiliser N rates in the most intensively farmed countries of northern Europe and eastern Asia were in the region of 120–550 kg N ha^{-1} (FAO, 1988). In contrast, the rates of total fertiliser applications in the countries of sub-Saharan African were in many cases less than 5% of those used in intensive agricultural systems. Between 1993/4 and 1997/8 world fertiliser nutrient usage increased by 13% (Isherwood, 2000).

An estimated 9.6 million tonnes of N fertiliser are applied to 140 million ha of agricultural land across Europe, of which half is used on wheat, barley, oats, maize and rye and a quarter on grassland (Aldinger, 2001). However, fertiliser N usage reached a peak in the mid to late 1980s and is expected to continue to decline to 2006 (by approximately 7% of the current total), from when it is thought that demand will remain stable (Aldinger, 2001).

Worldwide fertiliser N use has had an enormous impact upon agricultural productivity. It is difficult to separate out precisely the contribution of fertiliser N to increased output from that of the other technological inputs identified above by Wild (1993). Estimates in the UK (Hood, 1982) suggested that fertiliser N had been responsible for 30–50% of crop yield increases, with the remaining 50–70% due to improved varieties, increased agrochemical inputs and better husbandry techniques.

Increases in fertiliser N use have not occurred uniformly across all regions. In the UK, increased fertiliser use has tended to be concentrated in those areas of the country most suited to intensive agricultural production. These areas have seen significant changes in agricultural land use, notably a decline in traditional ley/arable rotational systems and an increase in continuous arable cropping. In much of eastern England agricultural land use is now dominated by intensive cereal production (Edwards and Withers, 1998).

Similar changes have been observed in many parts of the world as the availability of N fertiliser has helped to facilitate a profound change in cropping patterns. This has lead to a shift from mixed and multiple-cropping systems with relatively closed and self-sustaining N cycles to intensively managed monocultures with large N inputs in the form of synthetic fertilisers (Rosswall and Paustian, 1984). However, in some low-income developing countries this shift has also been welcomed, in that the cycle low-input low-output technologies that are perceived to perpetuate human drudgery with the ever present risk of hunger is broken (Isherwood, 2000).

With these increases in fertiliser usage and the accompanying changes in global cropping patterns, increasing attention has focussed on N leakage from the agroecosystem. Some concerns can be linked directly to fertiliser use, others relate to the overall intensification of modern agriculture resulting in part from the increased use of fertiliser N and the miscalculation of requirements for crop growth.

2.5 THE CAUSES OF NITRATE POLLUTION

The application of more N to a soil than can be assimilated by the soil or taken up by a crop creates a surplus of N. In the UK, in the early 1980s this surplus was in the order of 70 kg ha^{-1}, but is now down to 25 kg ha^{-1} (FMA, 1998). This surplus is particularly evident on dairy farms where inputs are often in the region of 400 kg N ha^{-1} but offtakes may only be in the order of 60–80 kg N ha^{-1} (Peel *et al.*, 1997). However, it is important to stress that even under well-managed arable land the N surplus may still be in the order of 20 kg ha^{-1} due to the mineralisation of organic N (Goulding, 2000).

The transfer and fate of this surplus has created many environmental concerns in Europe. The manifestations of the effect of the N surplus produced through agricultural production are considerable, especially as this surplus often behaves so conservatively in the soil system. The complexity of the behaviour of N in the agroecosystem has been illustrated in Figure 2.1 and with so many potential diffuse pathways for its loss from the agroecosystem, it is little surprise that the efficiency of applied N used by farmers never approaches 100%. Indeed it may even be as low as 10% in some grassland systems (Davies, 2000; Jarvis, 2000).

Over 80% of the total N in river waters is found in the form of NO_3^-, and in the last 30–40 years, NO_3^- levels in many European ground, surface and coastal waters have been gradually rising (House of Lords, 1990; DoE, 1986; Pau Vall and Vidal, 1999). An analysis of 12 UK rivers for which data was available over a 20-year period showed increases in concentration of between 50 and 400% (Wilkinson and Greene, 1982). Figure 2.6 shows the mean concentrations of NO_3^- in the Anglian region of eastern England from 1980 to 1998 and all are very close to the EU limit for NO_3^- in waters of 50 mg l^{-1} (DETR, 2001). The concentration of nitrate in rivers and reservoirs on the Channel island of Jersey have been found to consistently exceed 50 mg $NO_3^- l^{-1}$, due mainly to the large proportion of agricultural land that is devoted to the production of early potatoes. Losses from this crop can exceed 100 kg N ha^{-1} $year^{-1}$ (Lott *et al.*, 1999).

Rivers with the highest NO_3^- concentrations are found in the Midlands and south-east England, with the lowest levels in the mountainous regions of Wales, northern England and Scotland. The temporal and regional trends in river water quality are also reflected in lakes and reservoirs. For example, there has been a noticeable increase in reservoir NO_3-N levels in south-east England. Between 1992 and 1996 over 65% of European rivers had average annual NO_3^- concentrations greater than 1 mg l^{-1}, and of those 15% were greater than 7.5 mg l^{-1} (Pau Vall and Vidal, 1999).

Considerable quantities of freshwater are stored in groundwater aquifers, the most important of which are the chalk and Triassic sandstones. Long-term data on the NO_3-N concentration of groundwater are less common than for surface waters. Nevertheless there is a marked upward trend in many catchment areas (Figure 2.7), particularly in the dry eastern areas of England (Wilkinson and Greene, 1982; Davies, 2000; DETR, 2001). This is probably due to the fact that the NO_3^- concentrations in water draining from agricultural land are dependent upon both the amount leached but also the dilution rate. For example, catchments in the wetter west of the UK have considerable leaching (50 kg N ha^{-1} $year^{-1}$) but also greater rainfall, so reducing average NO_3^- concentrations (Davies, 2000).

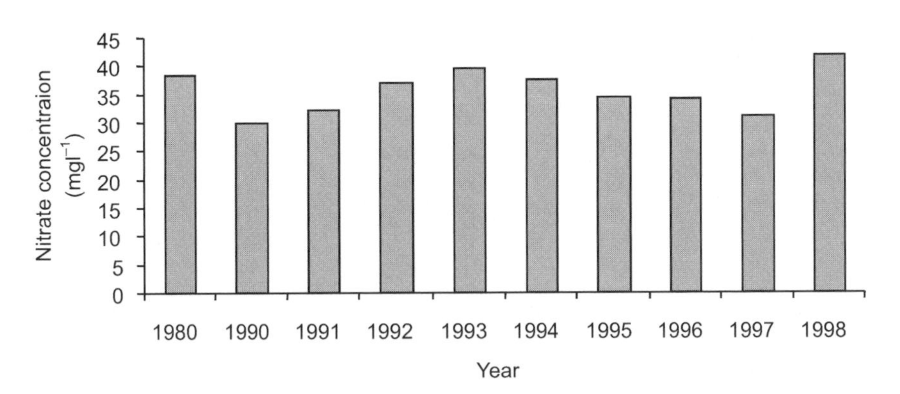

Figure 2.6 The mean concentration of NO_3^- (mg l^{-1}) in river waters from the Anglian region of Eastern England between 1980 and 1998 (DETR, 2001).

There is little doubt that the intensification of agriculture has been responsible for the increasing levels of NO_3^- in UK ground and surface waters (e.g. Royal Society, 1983; DoE, 1986; House of Lords, 1990; MAFF, 1993; Isherwood, 2000). It cannot be assumed, however, that this nitrate pollution is linked directly or solely to the increased use of N fertilisers; a correlation between the increase in NO_3^- pollution and the increase in fertiliser N use does not imply causality. On the contrary, NO_3^- loss from agricultural land is a complicated process that involves many factors.

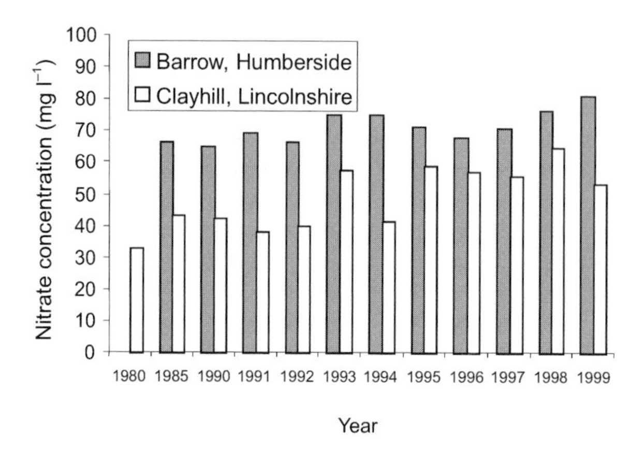

Figure 2.7 The change in mean NO_3^- concentrations (mg l^{-1}) from 1980 to1998 in two groundwater sources in Eastern England (DETR, 2001).

For example, intensive livestock production systems, produce large quantities of animal manure which can present disposal problems with excessive or untimely land application leading to both diffuse and/or point source NO_3^- pollution (Chapter 5). Furthermore, the surplus N in these systems can be considerable even when a low input minimal-loss strategy is taken (Figure 2.8) (Peel *et al.,* 1997).

The balance of the two major N inputs to agricultural soils, mineral fertilisers and livestock manures varies greatly across Europe, with the former being of greater importance in Denmark, Germany, Greece, France, Luxembourg, Finland and Sweden and the latter in Belgium and The Netherlands (Pau Vall and Vidal, 1999).

It is widely acknowledged that a major factor contributing to the increased pollution of the aquatic environment by NO_3^- has been the specialisation and intensification of agricultural enterprises (Chapter 3) (Edwards and Withers, 1998). Notably, the decline in traditional mixed farming systems (i.e. crops and livestock on the same farm) and the increase in specialist arable and livestock farms. These increase with the intensity of agricultural production and arise where N inputs (e.g. livestock feed) exceed N outputs (e.g. sales of crops and livestock). Nitrogen surpluses may be evident at a farm, regional or national level, although their occurrence and extent differs considerably between farming types and different countries. Specialist livestock farms (notably dairy, pigs and poultry) are particularly prone to accumulating

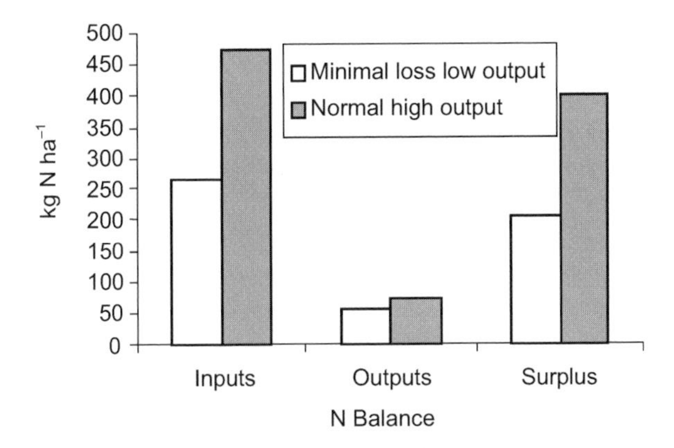

Figure 2.8 Measured N inputs, outputs and surpluses (kg N ha^{-1}) for two systems of dairy farm management in the Minimal Impact Dairy Systems (MIDAS1) experiment (Peel *et al.*, 1997).

excessive amounts of N in the form of manure because they use relatively high levels of N fertiliser and/or import large amounts of concentrate feed, and are commonly found in geographically concentrated areas (Chapter 5) (Lord *et al.*, 1999).

A specific trend observed in many areas of the UK has been the transition (stimulated largely by UK and subsequently EU agricultural policy) from pastoral and balanced rotational cropping to intensive arable production. The initial ploughing of grassland is a significant cause of NO_3^- leaching and elevated nitrate levels in groundwater have been specifically linked to the increased cultivation of permanent grassland to produce arable cropland during the 1940–1950s. Guidance exists to aid the farmer in taking into account the potential release of N following the ploughing of grass, which is dependent upon the amount of fertiliser N the grass received each year, the intensity of grass utilisation and time since ploughing (MAFF, 1994a). Importantly, it has also been established that even with this guidance it is possible to greatly over estimate N fertiliser requirements for cereal crops following grass leys. It is suggested that a reduction in N surplus would more likely be achieved if crop yield was removed as a determinant and an assessment of mineralisable soil N was made during the growing season (Withers and Sylvester-Bradley, 1999).

Furthermore, intensive arable cropping systems are intrinsically leaky (Powlson, 1988; MAFF, 1993) and prone to nitrate leaching since:

- they include periods of incomplete crop cover which are vulnerable to leaching events;
- regular cultivations stimulate N mineralisation and increase the NO_3^- availablity;

- repeated applications of fertiliser N increase the level of potentially mineralisable N in the soil, thereby increasing the amount of organically-derived NO_3^- that can be available for leaching;

- there is the risk of direct leaching losses from excessive (surplus due to poor N accounting) or poorly timed fertiliser N applications and organic manures (Richards *et al.,* 1999; Chambers *et al.,* 2000; Goulding, 2000; MAFF, 2000) (Chapter 5).

Nitrate pollution of ground, surface and marine waters is a major environmental issue in many European countries, with potential implications for human health, which has attracted considerable public and political attention. In contrast to pesticides, however, the control of NO_3^- pollution is not simply a question of rationalising and reducing an agrochemical input as outlined above. While the steady increase in the use of N fertilisers is undoubtedly significant in causing NO_3^- pollution, the overall relationship between the rate of N applied and the incidence of NO_3^- leaching is not always clear and direct. Nitrate leaching is caused by a number of factors and arises from a variety of sources, and this potentially complicates the control of NO_3^- pollution.

2.6 PROBLEMS CAUSED BY NITRATE POLLUTION

There are two key concerns regarding the NO_3^- pollution of water resources: the quality of drinking water and public health, and the eutrophication of surface waters. There is also some concern about increased levels of NO_3^- in foods and, although this is not strictly an agricultural pollution issue, it is given some consideration below.

Nitrate and drinking water quality

Much of Europe's drinking water is sourced from rivers and groundwaters. However, extensive leaching of NO_3^- from soils into these sources resulted in waters in many areas that approach or exceed the EU maximum mandatory limit concentration of 50 mg NO_3^- l^{-1}. An estimated 1 million people in the UK regularly drink water with greater than 50 mg $NO_3^-l^{-1}$ concentration (Packham, 1996).

Drinking water normally provides approximately up to 30% of the daily intake of NO_3^- ingested by humans, with the rest coming from fruit, vegetables and meat products (Isherwood, 2000). Nevertheless, the ingestion of large amounts of NO_3^- in drinking water may be harmful to humans since, although nitrate is relatively non-toxic and rapidly excreted from the body it can be reduced to potentially toxic nitrite in the mouth and gut (Magee, 1982). The main alleged health hazards of NO_3^- in drinking water are methaemoglobinaemia in young babies (<6 months old) and gastric cancer.

It has been known since the 1940s that excessive quantities of NO_3^- in drinking water may potentially present a health risk to young, bottle-fed babies (Fraser and Chilvers, 1981). Nitrite, derived from the bacterial reduction of ingested NO_3^-, is absorbed into the bloodstream where it combines with haemoglobin to form methaemoglobin that cannot transport oxygen. Globally there have been some 3000 cases of methaemoglobinaemia, commonly called 'blue baby syndrome' (Conway and Pretty, 1991). Most of these cases occurred prior to 1965 and were associated with bacterially contaminated private supplies or bottles. Cases still occur regularly in Hungary and Romania although the exact reason is not clearly known (Isherwood, 2000).

Under certain conditions, NO_3^- reacts with various amines to form nitrosamines and other N-nitroso compounds. Most of these compounds are strongly carcinogenic in animals (Magee, 1982) and there is evidence, for example, that N-nitroso compounds from tobacco can cause oral cancer (Hecht and Hoffmann, 1989). One site commonly regarded as being at risk from nitrosamines is the stomach and it has been suggested that an increase in human NO_3^- intake may lead to an increased risk of gastric cancer. However, no link to cancer (or diabetes in children) from this particular pathway has been demonstrated in humans (Isherwood, 2000; Wilson *et al.,* 1999). Indeed, some NO_3^- may even be beneficial in the reduction of potentially harmful pathogens in the human gut (Leifert *et al.,* 1999; Wilson *et al.,* 1999). For the moment, however, the gastric cancer debate is somewhat academic since international limits on NO_3^- levels have been implemented. The 1980 EC Directive on the Quality of Water Intended for Human Consumption, for example, set a 'maximum allowable concentration' of 50 mg NO_3^- l^{-1} and a guide level of 25 mg l^{-1}. The WHO reviewed this limit in 1997 in the light of increased scientific evidence on the topic, yet maintained the 50 mg NO_3^- l^{-1} (Isherwood, 2000).

Eutrophication

Many surface waters, such as rivers and lakes, have a limited supply of N and P and are described as oligotrophic (nutrient poor, low biological productivity). An environmental concern is that the enrichment of these waters with an available source of N and/or P can transform them into being eutrophic (nutrient rich, high biological activity). The most extreme examples of which are the dramatic algal blooms witnessed in some areas. The main problems associated with eutrophic waters are:

- changes in nutrient levels may affect the species composition of algal communities, with knock-on effects to other animals in the aquatic/marine food chain;
- decomposing algae and weeds can deoxygenate water causing fish mortality;
- algal blooms are often responsible for taints and problems with filtration in surface-drawn public water supplies, thereby increasing the cost of purification;
- surface algal blooms detract from the appearance of waters and impair their amenity value;
- some algal species produce ecologically-disruptive toxins which can also be hazardous to humans.

In most freshwaters in the UK, N is more abundant than P, yet the main cause of eutrophication is generally thought be increased levels of P (Chapter 3). However, opinions on this issue are conflicting (HCEC, 1987) and there are several instances where NO_3^- is considered to be the specific cause of freshwater eutrophication. Increasing concern is also being expressed about marine eutrophication in UK coastal and estuary waters, and it is thought that this may be specifically linked to N rather than P levels (Environment Agency, 2000).

Nitrates in food

Nitrate in drinking water is not the only source for consumers, since crops and ultimately food products may also contain significant quantities of NO_3^-, indeed in some cases this may be the main source of NO_3^- in the diet (RCEP, 1979).

Although NO_3^- is present in all crops, under some circumstances certain plants can accumulate very high concentrations. This is principally in the foliage e.g. spinach and cabbage, but may also occur in some storage roots (e.g. swedes and turnips) (Conway and Pretty, 1991). How much NO_3^- is in the harvested crop depends upon the amount of fertiliser N added, the crop characteristics and a variety of environmental factors, including season, sunlight and drought. The NO_3^- concentration of lettuce, for example, increases with increasing fertiliser N application and is higher in summer crops than those harvested in spring (Maynard *et al.,* 1976).

The full health implications of NO_3^- in food are again unclear and the overall picture is complicated by the NO_3^- intake from drinking water, as well as other dietary sources (e.g. some meats and cheeses) to which NO_3^- and nitrite are routinely added as preservatives. Claims that crops grown with the use of synthetic or mineral fertilisers have less taste and are not as nutritional beneficial as those that are organically grown has been shown to be incorrect (Woese *et al.,* 1995). While organically grown vegetables have higher DM contents and lower NO_3^- concentrations than those receiving synthetic fertilisers, there is little difference in regard to the parameters which affect sensory assessment and nutrition (Isherwood, 2000).

2.7 REDUCING NITRATE LOSSES FROM AGRICULTURE

Reduction in N fertiliser applications may be thought of as being the most straightforward method for effectively controlling N leakage from agriculture. However, this may not be the case, as profit is usually only made on the last few percent of the yield, and the reduction in N to below the economic optimum would sacrifice this profit. It has also been suggested that cutting fertiliser N inputs would have a limited impact on NO_3^- leaching in the short to medium-term (Davies, 2000). In the context of the EU and its CAP there is scope for developing a policy on fertiliser N restriction and reduce NO_3^- pollution which encompasses both the problems of N leakage and overproduction (DoE, 1986; MAFF, 1994a; MAFF, 1998b,c). To this end, interest has continued to be expressed in mechanisms such as N quotas and taxation (Clunies-Ross, 1993) (Chapter 9). For example in Denmark, the Netherlands and Norway fertiliser plans or schemes of nutrient accounting are now compulsory; if nutrient applications are over the maximum then fines may be imposed. In Denmark, it is also required to ensure that 65% of the cultivated area is covered with a green crop in winter, again to reduce NO_3^- loss (Isherwood, 2000).

Similarly, in England, in the early 1990s, 32 Nitrate Sensitive Areas (NSA) were designated in order to test practical measures for farmers to control NO_3^- loss. The areas were selected in regions in which groundwater sources had been shown to have high NO_3^- concentrations. The practical measures focussed on reducing the NO_3^- being generated or remaining in the soil in autumn and winter and maximising plant uptake of NO_3^- in autumn and early winter. The farmers were offered remuneration for adoption of measures being tested by the scheme and although there was considerable variation between individual areas, estimated reductions in NO_3^- losses were in the order of 30% (MAFF, 1994a; Lord *et al.*, 1999).

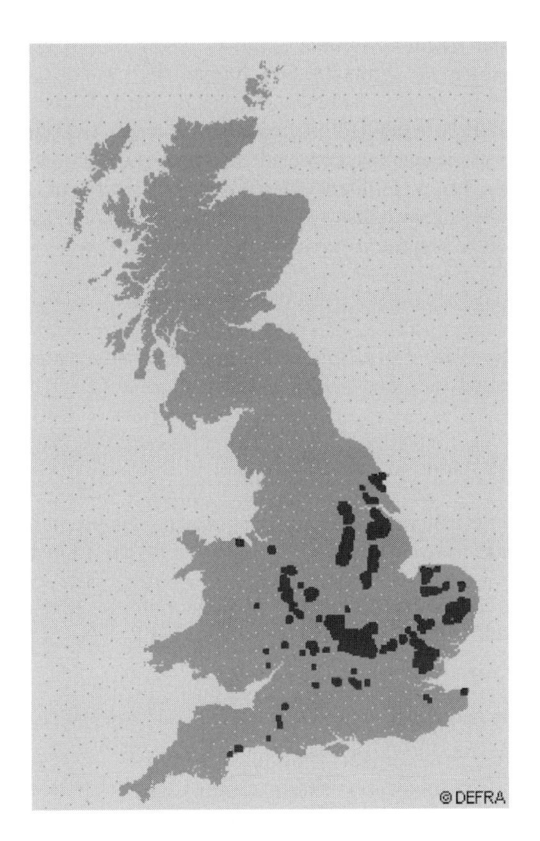

Figure 2.9 Location of nitrate vulnerable zones in England and Wales (DEFRA, 2001).

Further to the NSA scheme and in order to comply with the European Union Nitrate Directive, 68 Nitrate Vulnerable Zones (NVZs) were designated in the England and Wales in 1996 (Figure 2.9). In each of the zones, which in total cover an area of 600,000 ha, farmers are required to control fertiliser N inputs and ensure the appropriate and timely application of manure. An example of which may be the top dressing of manures to crops in spring rather than autumn (Dampney *et al.,* 2000). These zones are located in areas that are prone to NO_3^- leaching. For example, they may have sandy or course textured soils overlying unconfined aquifers that are significant sources of drinking water but are at particular risk of exceeding 50 mg NO_3^- l^{-1} (Davies, 2000). In contrast to the NSA scheme, the NVZ measures equate to good practice and so there are no financial incentives.

Significant control of N leakage from agriculture will only be achieved through the encouragement of management practices that minimise the opportunities for N (whether fertiliser-, soil- or manure-derived) to accumulate in a form that is susceptible to loss. Management practices that need to be encouraged in the UK and Europe to reduce NO_3^- leaching (especially during the autumn and winter months) include (MAFF, 1993, 1994b, 1998b,c; Chambers *et al.,* 2000; Dampney *et al.,* 2000; MAFF, 2000):

- ensuring fertiliser N is applied according to the crop's requirement, which depends upon the crop species/variety, expected yield and required quality; the soil N supply, including that released from soil organic matter, crop residues and applied manures (calculate fertiliser replacement values of manures (Box 2.2) and the extent to which available N will be lost before the crop takes it up;

- maximising usage of fertiliser technologies (Box 2.3);

- avoiding applications of N fertilisers and organic manures in autumn and very early spring when crop requirements are minimal;

- taking special care when applying fertilisers and organic manures on fields where there is a risk of run-off to surface waters;

- ensuring accurate calibration of fertiliser spreading equipment to minimise excessive application and ensure accurate and even placement (also for manures);

- using reduced cultivation techniques (e.g. direct drilling) rather than ploughing in the autumn to avoid enhancing soil N mineralisation and the production of excessive nitrate levels;

- minimising the period when the soil is left bare and susceptible to nitrate leaching by decreasing the area sown to spring crops and/or increasing the use of winter or cover crops (Box 2.4);

- sowing autumn crops early to increase nitrate uptake prior to the onset of the winter leaching period;

- restricting the ploughing of old grassland.

These practices already apply on voluntary basis to all UK farms as part of a comprehensive *Code of Good Agricultural Practice for the Protection of Water* (MAFF, 1998a) addressing all aspects of potential water pollution on the farm. This is a Statutory Code under the 1989 Water Act, which means that although contravention will not itself give rise to liability, failure to comply with the Code could be taken into account during any legal proceedings.

In the UK, the Department for the Environment, Food and Rural Affairs (DEFRA, formerly MAFF) currently invests £2.5 million per year on research and development programs studying the loss of NO_3^- from agricultural systems (MAFF, 2001). The European Fertiliser Manufacturers Association also provides guidance for usage of N fertiliser through codes of best agricultural practice for N, focusing on budgets and fertiliser plans, and urea targeting recommendation for effective use (EFMA, 2001).

The management strategies above fall into two categories: firstly, those that target the input or source of N into the agricultural system; and secondly, those that tackle reducing NO_3^- loss once the N is in the soil. Examples of how some of the strategies above may be put into practice and the expected results in regard to the reduction of NO_3^- leakage from agricultural systems are now considered. The results from many of the examples below have been incorporated into practical guidelines farmers can adopt to minimise the risk of NO_3^- pollution from their farming activities (Dampney *et al.*, 2000).

Source management – optimisation of fertiliser N use

The optimisation of fertiliser N usage is dependant upon adhering to the first 5 bullet points above. Central to this optimisation process is the calibration, upkeep and maintenance of manure and synthetic fertiliser spreaders is also crucial in the reduction of N loss, enabling accurate application and evenness of spread (Chapter 5). As well as calculation of the available N content of manures (Box 2.2) and the application rate, the timing of application, in regard to season and also preceding and forecast weather conditions is also crucial. Figure 2.10, emphasises the importance of timing in regard to manure application and the reduction of N leakage from a freely draining arable soil. In particular, it demonstrates the need to restrict autumnal applications, especially late in the season when rainfall is high and the manure contains large amounts of mineral N (as in this case, which is slurry/poultry manure).

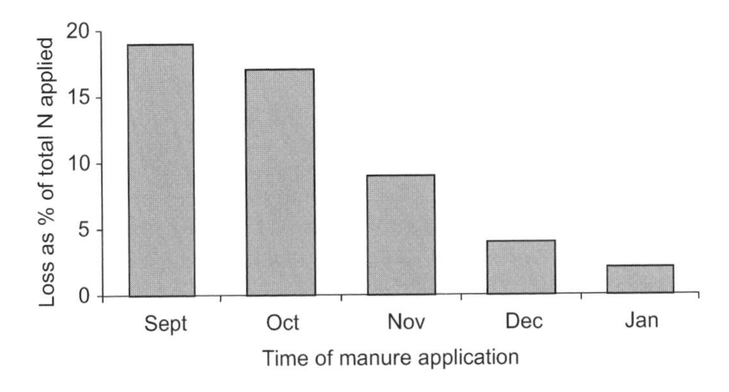

Figure 2.10 The variation in NO_3^- loss, expressed as a percentage of total N applied from manure (at 250 kg ha^{-1}) with month of application (Chambers *et al.*, 2000).

Box 2.2 Fertiliser replacement value of manures

The value of livestock manures as adjunct/alternatives to artificial inorganic fertilisers is well established, yet it still appears that some farmers have difficulty in determining their potential nutrient value, and so make little allowance for the nutrients supplied in manures in fertiliser plans (Chapter 5) (Owen, 1998).

Guidance on the fertiliser replacement value of manures can be calculated through a simple computer-based decision support program called the Manure Nitrogen Evaluation Routine (MANNER) which was developed through ADAS. Danish and Dutch models already exist but the required input information is somewhat complex. However, this model incorporates manure N analysis, NH_3 volatilisation, NO_3^- leaching and mineralisation of manure organic N (Chambers *et al.,* 1999). Validation of the model outputs against independently collected data from a range of manures spread onto a variety of arable treatments has given very good agreement (Chambers *et al.,* 1999).

Cultivation and crop management

As cultivation stimulates N mineralisation, manipulation of the timing and practice of cultivation has a key role to play in control of nitrate loss from agricultural land. Some of the key factors are incorporated into recommendations for NVZs, and include:

- Minimising the opportunity for N mineralisation by reduced cultivation techniques (e.g. direct drilling) rather than ploughing in the autumn. Experiments conducted at Brimstone Farm, Oxfordshire during the 1980s confirmed that ploughing and secondary cultivations increased the loss of nitrate by 21% compared with direct drilling (Goss *et al.,* 1994). Average losses from winter wheat over an 8-year period were 30 kg N ha^{-1} from the ploughed soil and 23 kg N ha^{-1} from the direct drilled soil. If cultivation is necessary to establish a uniform seedbed, then losses can be minimised by shallow ploughing or using a non-inversion rigid tine cultivator or heavy discs as primary cultivation (Johnson *et al.,* 1997).

- The length of time that the soil is left bare and susceptible to nitrate leaching after the harvest of the previous crop has also been found to be important in controlling nitrate losses. Goss *et al.* (1994) observed that the fate of the decomposing residues from the previous crop exerted a strong control on the magnitude of nitrate losses in arable soils. Hence, sowing autumn crops early will increase nitrate uptake prior to the onset of the winter leaching period.

- Restricting the ploughing of old grassland careful management of residues. Grass crops can leave large amounts of residual plant material in or on the soil are potentially considerable sources of N leakage. Residues from crops such as potatoes, peas and oil seeds (and also ley crops) need to be managed and considered in regard to fertiliser plans (Adams and Jan, 1999). Simultaneous sowing a cover crop and lifting potatoes ('liftsowing') has been shown to increase the amount of N in the above ground plant material and reduce the mineralised N in the soil compared with a conventionally drilled cover crops (Box 2.4) (Buchner *et al.*, 1997).

Box 2.3 Fertiliser products limiting N loss

Improvements in the utilisation of fertilisers and reductions in the N surplus in agroecosystems are often largely attributable to changes in agricultural practice, fertiliser application techniques or crop plant manipulation (Isherwood, 2000). There has been very little change in the basic fertiliser product in many years, which is thought to the lack of incentive to invest in the development of a bulky low-priced commodity that offers limited scope for product differentiation (Isherwood, 2000).

Slow release or slowly available N compounds

Slow release N fertilisers extend the period over which N is released into the soil compared to the conventional highly water soluble rapidly available fertilisers. The advantages of this are that it may be possible to match the requirements of the crop and reduce the need for repeated fertiliser applications through the growing season (Tisdale *et al.,* 1993).

Nitrogen containing fertiliser compounds that are sparingly soluble (or requiring microbial action for release on reaching the soil) and so slower release sources, in comparison to conventional fertilisers include:

- ureaforms, a generic term for a range of materials which are mixtures of methylene ureas, these may have approximately 40% N in a relatively insoluble form;
- coated urea, in which the solubilisation of the N-containing urea is reduced by coating with sulphur or acrylic waxes;
- sparingly soluble simple compounds such as magnesium ammonium phosphate, in which solubility is largely controlled by particle size.

These fertilisers tend to be expensive and so their use is often restricted to high-yielding cash crops and not broad-scale agriculture. They are especially useful when used on irrigated sandy soils from which leaching losses would be expected to be high (Tisdale *et al.,* 1993).

A further approach to the slow release of N is to maximise efficiency and reduce losses by actively attempting to inhibit the micro-organisms responsible for the nitrification reaction (the conversion of NH_4^+ to NO_3^-) in soil. The period of inhibition is temporary and losses of N will be restricted only if the period of maximum loss (e.g. excessive rainfall) coincides with the period of inhibition. One of the most common inhibitors is N-Serve or nitrapyrin, acts by inhibiting cytochrome oxidase involved in the oxidation of NH_4^+ by *Nitrosamonas*. Concentrations of 0.1–20 mg nitrapyrin kg^{-1} soil are generally required to produce an effect that lasts for 4–6 weeks. Improvements in fertiliser efficiency and reduction in N loss have been observed using these methods. Again, expense of testing and subsequent technology transfer to farmers are given as reasons limited adoption of this alternative.

Other, even less widely used alternatives are nutrient absorption enhancers and microbial inoculants. Absorption enhancers basically restrict the plant availability (and solubility) of added N through absorption to polymers. Microbial seed inoculants are used to enhance the biological N fixation and to supplement the synthetic fertilisers. For example, the inoculation of *Rhizobium* can be beneficial to leguminous plants, such as peas and beans (Isherwood, 2000).

Returning arable land to low-input grassland grassland has been shown to be one of the most effective strategies in reducing NO_3^- leaching (Lord *et al.*, 1999). However, this may not be economically possible for many farmers, although rotational grazing of livestock is an option. Unsurprisingly, increased stocking density and fertiliser rates increase NO_3^- losses, but it has been suggested that through the use of clover-based swards (that biologically fix atmospheric N) and cutting inputs, losses may be greatly reduced. However, the evidence for this is mixed, and of greater importance is probably the number of livestock grazing days spent in the field which closely relates to N inputs from animal excreta (Cuttle *et al.*, 1998). Attempts to counter this input by and reducing fertiliser N inputs, both spatially and temporally and so reduce NO_3^- losses are technically possible, although probably prohibitively expensive (Cuttle *et al.*, 2001).

Finally, it is important to realise that one of the most important factors that determine loss and impact of NO_3^- from agricultural systems is the weather. While this may be beyond the control of the farmer, it can nevertheless be taken into consideration. Temperature and rainfall have a large bearing on drainage and crop growth (and or course N mineralisation). This has been clearly observed by Adams and Jan (1999) who noted that rainfall forecasts 4–6 weeks in advance and calculations of soil moisture deficit could be used to maximise the utilisation of biologically fixed N for autumn sown crops and therefore limit N leakage in the form of NO_3^- loss.

It is a widely held view that whilst some of the above strategies may, when practised in isolation, have a modest impact on N loss, it is through the adoption of a combination of these practical measures that losses will be reduced (Peel *et al.*, 1997).

Box 2.4 Cover crops

As crops are being harvested during the summer the soil beneath them is accumulating mineral N, in particular NO_3^-, derived from the microbial breakdown of soil organic matter. In autumn, still more NO_3^- will be produced which will eventually be leached from the soil in drainage water by winter rain. Growing a cover crop during this 'leaky' stage in a rotation 'soaks-up' much of the excess NO_3^- and also increases rates of evapotranspiration, reducing drainage. Furthermore, the valuable N absorbed is effectively stored for re-release at a time when, upon incorporation and decomposition of the cover crop, it can be used by subsequent crops or used as fodder. The effect of a cover crop grown before spring crops (and after barley) on reducing NO_3^- leachate in winter, compared to winter sown cereals is shown in the figure below (Lord *et al.,* 1999).

Research suggests that an early-sown and well-established autumn crop can reduce N losses by between 20 and 50 kg ha^{-1} $year^{-1}$. However, this may be variable and is dependent upon initial establishment of the cover crop (Lord *et al.,* 1999). In a 3-year study Harrison *et al.* (1996) observed that winter rye was most effective in uptake of N when compared to white mustard, oilseed rape, turnips and winter barley, yet still varied between 8 and 27 kg ha^{-1} $year^{-1}$. This was thought to be attributable to weather and soil conditions.

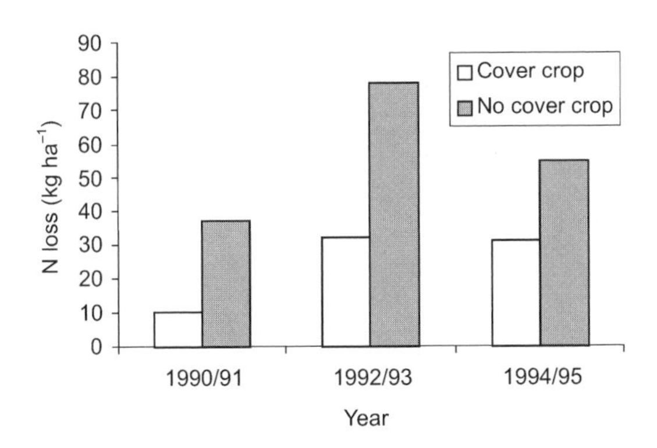

The principle objective of sowing a cover crop must be to establish soil cover as soon as possible, thereby obtaining the maximum autumn growth. Ideally, cover should be established and actively growing by mid-September; sowing any later than mid-October is likely to have little effect on NO_3^- leaching. It has also been suggested that by delaying cultivation to this time (when the soil is colder and so mineralisation of N slower) may also result in reduced N loss (Johnson *et al.,* 1997).

In Sweden, undersown catch crops, such as a grass crop sown with spring cereals, have been shown to be particularly effective in limiting N leakage. Reductions of 40–50% of leached N were measured from beneath the plots sown with the catch crop when compared with the control tillage treatment. The main crop is harvested at the normal time, but the catch crop remains covering the soil until the following spring. Importantly, the time of sowing is crucial so as to ensure that the main crop does not suffer adversely from competition effects from the catch crop (Aronsson and Torstensson, 1998). A drawback is that weed, pest and disease problems are thought to be more prevalent when cover and catch crops are used (Goulding, 2000).

To conclude, agricultural land is the main source of nitrate in most UK groundwaters and rivers. Extensive research has been conducted throughout Europe over the last three decades, such that the relationship between nitrate inputs, outputs and transformations is now well understood. The EU Nitrates Directive, adopted in 1991, places the onus on

Member States to reduce nitrate pollution by introducing controls on agricultural practices where the concentration of nitrate in water exceeds, or is risk of exceeding 50 mg $NO_3\,l^{-1}$. Guidance is available from many sources, for example DEFRA and the EA, to demonstrate the most effective way of reducing nitrate losses in water draining from agricultural land. Current legislation has targeted vulnerable catchments that have been designated as NVZs. In these areas, farmers are required to change their practice to reduce pollution risk, with no direct compensation. The NSAs scheme, which provides payment for loss of income associated with changes in practice, comes to an end in 2005. The future extension of schemes to limit nitrate losses from agricultural land will depend on the success of the NVZ scheme and the on-going debate surrounding the health risks associated with the supply of drinking water containing elevated nitrate concentrations.

Technological developments will play a major role in the continuing advances made to enhance the efficiency of nitrogen use in agriculture. For example, targeting N fertiliser inputs in cereal systems through the application of precision farming techniques offers the opportunity to further reduce nitrate losses. Considerable investment in global positioning system equipment and yield monitoring combine harvesters, together with a detailed programme of soil mineral nitrogen analyses is necessary to provide the information from which variable rate applications of fertiliser and other inputs can be made (Welsh *et al.,* 1999; Taylor *et al.,* 2000). Research has demonstrated that the main factors influencing the optimum N rate are soil type and supply of N from the soil. Further work is on-going across Europe to assess the condition of the cover canopy during the season, with the aim of utilising variable rate application technology to adjust inputs to optimum levels. The techniques look promising. For example, AGCO/ADAS (1999) observed that adjusting N inputs based on variations in soil mineral N assessment within fields lead to lower fertiliser N inputs and reduced soil N levels in residues while maintaining yields at levels equivalent to conventional winter wheat crops.

The net effect of the increased uptake of new technology, together with the adoption of nutrient-conserving practices based on more traditional approaches will be a continued reduction of N losses from agricultural land. In parallel with these environmentally beneficial changes in farming practice, economic benefits will accrue to individual farmers due to the more efficient use of inputs, machinery and farm labour.

REFERENCES

Adams, W.A. and Jan, M.T. (1999) Utilization of nitrogen accumulated by a clover containing ley following cultivation. *Soil Use and Management* **15**, 247–253.

Addiscott, T. (1996) Fertilisers and nitrate leaching. In: *Issues in Environmental Science and Technology 5: Agricultural Chemicals and the Environment* (Eds R.E. Hester and R.M. Harrison). The Royal Society of Chemistry, Letchworth, pp. 1–26.

AGCO/ADAS (1999) Yield Mapping and Precision Farming: an Appraisal of Potential Benefits on Recent Research and Farmer Experience. AGCO, Coventry, pp. 17.

Aldinger, H. (2001) *The Fertiliser Sector in Western Europe, Market Trends and Industry Outlook.* Presented to the 2nd International Conference on the Chemical and Petrochemical Industries of Russia the CIS and Central and Eastern Europe. Budapest, 23–24th April. European Fertiliser Manufacturers Association, Brussels.

Arah, J.R.M., Smith, K.A., Crichton, I.J. and Hesheng, L. (1991) Nitrous oxide production and denitrification in Scottish soils. *Journal of Soil Science* **42**, 351–367.

Aronsson, H. and Torstensson, G. (1998) Measured and simulated availability and leaching of nitrogen associated with frequent use of cash crops. *Soil Use and Management* **14**, 6–13.

Asman, W.A.H., Sutton, M.A. and Schjørring, J.K. (1998) Ammonia: emission, atmospheric transport and deposition. *New Phytologist* **139**, 27–48.

Brady, N.C. and Weil, R.R. (1999) *The Nature and Properties of Soil.* 12th Edition. Prentice Hall, New Jersey.

Buchner, W., Knechtges, H. and Sroka, J. (1997) Liftsowing' at the potato harvest. For a faster uptake of nitrogen by cover crops. *Kartoffelbau* **48**, 308–310.

Chambers, B.J., Lord, E.I., Nicholson, F.A. and Smith, K.A. (1999) Predicting nitrogen availability and losses following application of organic manures to arable land: MANNER. *Soil Use and Management* **15**, 137–143.

Chambers, B.J., Smith, K.A. and Pain, B.F. (2000) Strategies to encourage better use of nitrogen in animal manures. *Soil Use and Management* **16**, 157–161.

Clunies-Ross, T. (1993) Taxing nitrogen fertilisers. *The Ecologist* **23**, 13–17.

Conway, G.R. and Pretty, J.N. (1991) *Unwelcome Harvest: Agriculture and Pollution.* Earthscan, London.

Cuttle, S.P., Scurlock, R.V. and Davies, B.M.S. (1998) A 6-year comparison of nitrate leaching from grass/clover and N-fertilized grass pastures grazed by sheep. *Journal of Agricultural Science* **131**, 39–50.

Cuttle, S.P., Scurlock, R.V. and Davies, B.M.S. (2001) Comparison of fertiliser strategies for reducing nitrate leaching from grazed grassland, with particular reference to the contribution from urine patches. *Journal of Agricultural Science* **136**, 221–230.

Dampney, P.M.R., Lord, E.I. and Chambers, B.J. (2000) Development of improved advice for farmers and advisers. *Soil Use and Management* **16**, 162–166.

Davies, D.B. (2000) The nitrate issue in England and Wales. *Soil Use and Management* **16**, 142–144.

De Klein, C.A.M., Sherlock, R.R., Cameron, K.C. and Van der Weerden, T.J. (2001) Nitrous oxide emissions from agricultural soils in New Zealand – a review of current knowledge and directions for future research. *Journal of The Royal Society of New Zealand* (in press).

DoE (1986) *Nitrate in Water: a report by the Nitrate Co-ordination Group.* Department of the Environment Pollution, HMSO, London, Paper No. 26.

DETR (2001) Chapter 3 Inland Water Quality and Use. In: *Digest of Environmental Statistics.* Department of the Environment, Transport and the Regions, London.

Edwards, A.C. and Withers, P.J.A. (1998) Soil phosphorus management and water quality: a UK perspective. *Soil Use and Management* **14**, 124–130.

EFMA (2001) *Sustainable Soil Management an Achievable Goal.* European Fertiliser Manufacturers Association, Brussels.

Environment Agency (2000) *Aquatic Eutrophication in England and Wales: a Management Strategy.* Department of the Environment, Transport and the Regions.

FAO (1988) *Fertiliser Yearbook Volume 37.* Food and Agriculture Organisation, Rome.

FMA (1998) *The Fertiliser Review 1998.* Fertiliser Manufacturers Association, Peterbourough.

FMA (2000) *The Fertiliser Review 2000.* Fertiliser Manufacturers Association, Peterbourough.

Folorunso, O.A. and Rolston, D.E. (1984) Spatial variability of field-measured denitrification gas fluxes. *Soil Society of America Proceedings* **48**, 1214–1219.

Fowler, D., Sutton, M.A., Skiba, U. and Hargreaves, K.J. (1996) Agricultural nitrogen and emissions to the atmosphere. In: *Issues in Environmental Science and Technology 5: Agricultural Chemicals and the Environment* (Eds R.E. Hester and R.M. Harrison). The Royal Society of Chemistry, Letchworth, pp. 57–84.

Fraser, P. and Chilvers, C. (1981) Health aspects of nitrate in drinking water. *Science of the Total Environment* **18**, 103–116.

Goss, M.J., Howse, K.R., Lane, P.W., Christian, D.G. and Harris, G.L. (1994) Losses of nitrate-nitrogen in water draining from under autumn-sown crops established by direct drilling or mouldboard ploughing. *Journal of Soil Science* **44**, 35–48.

Goulding, K.W.T., Bailey, N.J., Bradbury, N.J., Hargreaves, P., Howe, M., Murphy, D.V., Poulton, P.R. and Willison, T.W. (1998) Nitrogen deposition and its contribution to nitrogen cycling and associated soil processes. *New Phytologist* **139**, 49–58.

Goulding, K. (2000) Nitrate leaching from arable and horticultural land. *Soil Use and Management* **16**, 145–151.

Harrison, R., Davies, D.B., Peel, S., van Cleemput, O., Hofman, G. and Vermoesen, A. (1996) Nitrogen uptake by cover crops at 7 sites in the UK: 1990–1993. *Developments in Plant and Soil Sciences* **68**, 203–205.

Haygarth, P.M., Chapman, P.J., Jarvis, S.C. and Smith, R.V. (1998) Phosphorus budget for two contrasting grassland farming systems in the UK. *Soil Use and Management* **14**, 160–167.

HCEC (1987) *Pollution of rivers and estuaries, Vol. II: Minutes of evidence and appendices.* House of Commons Environment Committee, Third Report, Session 1986–87. HMSO, London.

Hecht, S.S. and Hoffmann, D. (1989) The relevance of tobacco and specific nitrosamines to human cancer. *Cancer Surveys* **8**, 273–294.

Hood, A.E.M. (1982) *Fertiliser Trends in Relation to Biological Productivity Within the UK.* Philosophical Transactions of the Royal Society, London, B 296, 315–328.

House of Lords (1990) *Nitrate in Water.* 16th Report of the Select Committee on the European Communities (HL Paper 73). HMSO, London.

Isherwood, K.F. (2000) *Fertiliser Use and the Environment.* (Revised Edition). International Fertiliser Industry Association. Paris.

Jansson, S.L. and Persson, J. (1982) Mineralisation and immobilisation of soil nitrogen. In: *Nitrogen in Agricultural Soils* (Ed. Stevenson, F.J.), Agronomy Monograph No. 22, American Society of Agronomy, Madison, Wisconsin, pp. 229–252.

Jarvis, S.C. (2000) Progress in studies of nitrate leaching from grassland soils. *Soil Use and Management* **16**, 152–156.

Jenkinson, D.S. (1990) An introduction to the global nitrogen cycle. *Soil Use and Management* **6**, 56–60.

Johnson, P.A., Shepherd, M.A. and Smith, P.N. (1997) The effects of crop husbandry on nitrate leaching from a shallow limestone soil growing a five course combinable crop rotation. *Soil Use and Management* **15**, 17–23.

Laity, J. (1948) *Profitable Ley Farming.* Crosby Lockwood & Son Ltd, London.

LaRue, T.A. and Patterson, T.G. (1981) How much nitrogen do legumes fix? *Advances in Agronomy* **34**, 15–38.

Leifert, C., Fite, A., Hong, Li., Golden, M., Mowet, A. and Frazer, A. (1999) Human health effects of nitrate. IFA Conference on managing plant nutrition. Barcelona, July 1999.

Lord, E.I., Johnson, P.A. and Archer, J.R. (1999) Nitrate sensitive areas: a study of large scale control of nitrate loss in England. *Soil Use and Management* **15**, 201–207.

Lord, E.I. and Anthony, S.G. (2000) MAGPIE: A modeliing framework for evaluating nitrate losses at national and catchment scales. *Soil Use and Management* **16**, 167–174.

Lott, J., Parkinson, R.J., Fuller, M.P. and Beattie, L. (1999). Evaluating strategies to reduce nitrate and phosphate loss to surface water from agricultural land on Jersey. *International Association of Hydrological Sciences Publication* **257**, 221–228.

Mackenzie, G.H. and Traureau, J.C. (1997) *Recommendation Systems for Nitrogen.* The Fertiliser Society. Proceedings No. 403.

MAFF (1993) Solving the Nitrate Problem: progress in research and development. Publication No. 1092, Ministry of Agriculture, Fisheries and Food Publications, London.

MAFF (1994a) *The Nitrate Sensitive Areas Scheme.* Ministry of Agriculture, Fisheries and Food, London.

MAFF (1994b) *Fertiliser Recommendations for Agricultural and Horticultural Crops.* Ministry of Agriculture, Fisheries and Food Reference Book 209. HMSO, London.

MAFF (1998a) *Code of Good Agricultural Practice for the Protection of Water.* Revised 1998. Ministry of Agriculture, Fisheries and Food, London.

MAFF (1998b) *Guidelines for Farmers in Nitrate Vulnerable Zones.* Ministry of Agriculture, Fisheries and Food, London.

MAFF (1998c) *Manure Planning in Nitrate Vulnerable Zones.* Ministry of Agriculture, Fisheries and Food, London.

MAFF (2000) *Towards Sustainable Agriculture: Pilot Set of Indicators.* Ministry of Agriculture, Fisheries and Food, London.

MAFF (2001) Factors affecting the fate of organic nitrogen. *Environmental R+D Newsletter* Ministry of Agriculture, Fisheries and Foods, London. No. 7, p. 8.

Magee, P.N. (1982) Nitrogen as a potential health hazard. *Philosophical Transactions of the Royal Society* London B 296, 543–550.

Maynard, D.N., Barker, A.V., Minotti, P.L. and Peck, N.H. (1976) Nitrate accumulation in vegetables. *Agronomy Journal* **28**, 71–118.

Murphy, D.V., MacDonald, A.J., Stockdale, E.A., Goulding, K.W.T., Fortune, S., Gaunt, J.L., Poulton, P.R., Wakefield, J.A., Webster, C.P. and Wilmer, W.S. (2000) Soluble organic nitrogen in agricultural soils. *Biological Fertility of Soils* **30**, 374–387.

Owen, L. (1998) *Use of Organic Manure; Storage of Liquid Fertilisers*. The British Survey of Fertiliser Practice, 1994, Edinburgh University Data Library. pp. 18.

Packham, R.F. (1996) Drinking water quality and health. In: *Pollution, Causes, Effects and Control* (Ed. R.M. Harrison). 3rd Edition. The Royal Society of Chemistry, Cambridge, pp. 52–65.

Pau Vall, M. and Vidal, C. (1999) Nitrogen in agriculture. In: *Agriculture, Environment, Rural Development: Facts and Figures – A Challenge for Agriculture*. European Union, Brussels.

Parsons, J.W. (1984) Green manuring. *Outlook on Agriculture* **13**, 20–23.

Peel, S., Chambers, B.J., Harrison, R. and Jarvis, S.C. (1997) Reducing nitrogen emssions from complete dairy farm systems. In: *Gaseous Nitrogen Emissions from Grasslands* (Eds S.C. Jarvis and B.F. Pain). CAB International, Wallingford, pp. 383–390.

Powlson, D.S. (1988) Measuring and minimising losses of fertiliser nitrogen in arable agriculture. In: *Nitrogen Efficiency in Agricultural Soils* (Eds D.S. Jenkinson and K.A. Smith). Elsevier, London, pp. 231–245.

Powlson, D.S. (1993) Understanding the soil nitrogen cycle. *Soil Use and Management* **9**, 86–94.

Powlson, D.S. (2000) Tackling nitrate from agriculture. *Soil Use and Management* **16**, 141.

Rahn, C.R., Paterson, C.D., Willison, T.W. and Lillywhite, R. (1996) The fate of nitrogen from brassica crop residues in intensive cropping rotations. In: *Proceedings of COST 814 Conference on 'Nitrogen supply and nitrogen fixation of crops for cool wet climates'*. Trømso, Norway. European Commission, Brussels, pp. 219–226.

Richards, I.R., Turner, I.D.S. and Wallace, P.A. (1999) Manure and fertiliser contributions to soil mineral nitrogen and the yield of forage maize. *Nutrient Cycling in Agroecosystems* **55**, 175–185.

RCEP (1979) *Agriculture and Pollution. 7th Report of the Royal Commission on Environmental Pollution* (Cmnd 7644). HMSO, London.

RCEP (1996) *Sustainable Use of Soil*. Royal Commission on Environmental Pollution 19th Report (Cm 3165). HMSO, London.

Rosswall, T. and Paustian, K. (1984). Cycling of nitrogen in modern agricultural systems. *Plant and Soil* **76**, 3–21.

Rowell, D.L. (1994) *Soil Science, Methods and Applications*. Longman Scientific and Technical, Harlow.

Royal Society (1983) *The Nitrogen Cycle of the United Kingdom: a Study Group Report*. Royal Society, London.

Schlesinger, W.H. (1997) *Biogeochemistry, an Analysis of Global Change*. 2nd Edition. Academic Press, San Diego.

Simmelsgaard, S.E. (1999) The effect of crop, N-level, soil type and drainage on nitrate leaching from Danish soil. *Soil Use and Management* **14**, 30–36.

Smith, K.A. and Arah, J.R.M. (1990) *Losses of Nitrogen by Denitrification and Emissions of Nitrogen Oxides from Soils*. Proceedings No. 299, The Fertiliser Society, London.

Speirs, R.B. and Frost, C.A. (1987) The enhanced acidification of a field soil by very low concentrations of atmospheric ammonia. *Research and Development in Agriculture* **4**, 83–86.

Taylor, J.C., Wood, G.A., Welsh, J.P. and Knight, S. (2000) Exploring management strategies for precision farming of cereals assisted by remote sensing, aspects of applied biology. *Remote Sensing in Agriculture* **60**, 53–60.

Tisdale, S.L, Nelson, W.L., Beaton, J.D. and Halvin, J.L (1993) *Soil Fertility and Fertilisers*. 5th Edition. Prentice Hall, New Jersey.

Welsh, J.P., Wood, G.A., Godwin, R.J., Taylor, J.C., Earl, R., Blackmore, B.S., Spoor, G. and Thomas, G. (1999) Developing strategies for spatially variable nitrogen application. In: *Precision*

Agriculture '99 (Ed. J. Stafford). Proceedings of the 2nd European Conference on Precision Agriculture, Odense Congress Centre, Denmark, pp. 729–738.

White, R.E. (1997) *Principles and Practice of Soil Science, the Soil as a Natural Resource*. 3rd Edition. Blackwell Science Ltd, Oxford.

Whitehead, D.C., Pain, B.F. and Ryden, J.C. (1986) Nitrogen in UK grassland agriculture. *Journal of the Royal Agricultural Society of England* **147**, 190–201.

Whitehead, D.C. (2000) *Nutrient Elements in Grassland: Soil–Plant–Animal Relationships*. CABI Publishing, Wallingford.

Wild, A. (1988) Plant nutrients in soil: nitrogen. In: *Russell's Soil Conditions and Plant Growth* (11th Edition) (Ed. A. Wild), Longman, Harlow, pp. 652–694.

Wild, A. (1993) *Soils and the Environment: an Introduction*. Cambridge University Press, Cambridge.

Wilkinson, W.B. and Greene, L.A. (1982) The water industry and the nitrogen cycle. *Philosophical Transactions of the Royal Society London* **B 296**, 459–475.

Wilson, W.S., Ball, A.S. and Hinton, R.H. (Eds) (1999) Managing the risk of nitrates to humans and the environment. *Proceedings of a Masterclass Conference Organised by the Agricultural Sector and Toxicology Group of the Royal Society of Chemistry*. The Royal Society of Chemistry, Cambridge.

Withers, P.J.A. and Sylvester-Bradley, R. (1999) Nitrogen fertiliser requirements of cereals following grass. *Soil Use and Management* **15**, 221–229.

Woese, K., Lange, D., Boess, C. and Bögl, K.W. (1995) Ökologisch und konventionell erzeugte Lebensmittel im Vergleich. Eine Literaturstudie. *Bundesinsitut für gesundheitlichen Verbraucherschutz und Verterinärmedizin* **4**.

Phosphorus 3

3.1 INTRODUCTION

Before the 1980s research into phosphorus (P) in agro-ecosystems focussed on crop growth, yield and quality. Contemporary interest in the environmental behaviour of P is now largely concentrated on its movement to rivers, lakes and seas. It was thought that, unlike nitrates, phosphates were held so strongly within the soil matrix that movement to natural waters was negligible. However, the increased incidence of eutrophic waters, algal blooms and other ecologically damaging effects for which P is a major contributing factor, indicate this not to be the case. Agricultural land is now recognised to be the largest non-point source of P reaching surface waters. Morse *et al.* (1993) estimate that 43% of the phosphorus inputs into surface waters in the UK are attributable to agriculture. The build-up of P levels in soils over the last 45 years also had an impact upon plant species diversity in grassland meadows and semi-natural habitats. This chapter addresses:

- the reasons for the increase in P movement within the environment;
- the potential environmental and human health implications of this increase;
- strategies to minimise P losses and maximise usage within agroecosystems.

Phosphorus and agriculture

Phosphorus is a macronutrient that is essential to all forms of life. The most essential functions of P in organisms are in energy storage and transfer and as a component of DNA, vital for the passing on of heredity traits from one generation to another. Phosphorus is only found in combination with other elements due to the reactive nature of elemental P.

Phosphorus, water and N tend to be the dominant yield-limiting factors for agricultural crop growth. Concentrations of P in plants are normally in the range of 0.1–0.4% (expressed by weight on a DM basis), which is very much lower than those of N (1.5%) and K (1%). However, there is considerable variation in P content between species, with seed bearing and root crops having a greater P demand. There is also considerable variation in P content between parts of the same plant, with seeds and grain showing higher contents than straw, stalks and stover (Table 3.1).

Because of the involvement of P in energy transfer reactions, it is no surprise that it plays a role in plant photosynthesis, respiration, cell enlargement and division. Indeed, almost every metabolic reaction of any significance proceeds via a phosphate derivative. Phosphorus also promotes early root formation and growth, and as plants mature most P moves into seeds and/or fruiting bodies, hence the quality of grain, fruit and vegetable crops is greatly improved with an adequate supply (Tisdale *et al.,* 1993). A decrease in root disease and an increase in cold tolerance in small grains and improved cereal stalk strength has also been observed under conditions of optimum P supply.

Table 3.1 Phosphorus content of plant parts (Glendinning, 1999).

Crop Plant	Plant part	Percentage P content
Soybeans	Grain	0.42
	Straw	0.18
Maize	Grain	0.22
	Stover	0.17
Cotton	Seed	0.66
	Stalks	0.24
Oilseed Rape	Grain	0.65
	Stalks	na
Wheat	Grain	0.42
	Straw	0.12

The dominant feature of soil P is its strong adsorption to soil particle surfaces – giving very low solution concentrations and the very low solubility of P minerals ($<0.01–1 \text{ mg l}^{-1}$). As a consequence crop plant deficiencies of P are common (Rowell, 1994). The greatest concentrations of P in young plants are in tissue at the growing point, and as P tends to be redistributed from old tissue to new, a deficiency is often observed in the lower parts of plants. However, unlike N and K, visual foliar deficiency symptoms are not obvious, due to the reduced maturity and stunted nature of crops grown under P deficient conditions, detection may be very difficult. However, in some cereal and grass species a purple or red tinge may appear on older leaves caused by an accumulation of sugars, or at certain growth stages the crop may have leaves which are much darker green than normal (Glendinning, 1999). Phosphorus toxicity, while not common, may occur at excessive P concentrations appearing as interveinal chlorosis and necrosis in younger leaves as well as the shedding of older leaves (Reuter and Robinson, 1997).

The maintenance of adequate soil P levels are essential for the development and production of agricultural crops. Hence, P application to soils as a fertiliser is a necessity in modern farming systems. Before consideration of the transfer and movement of P from agriculture into the wider environment, it is essential to consider those factors influencing the occurrence, form and behaviour of P within agroecosystems.

3.2 THE PHOSPHORUS CYCLE

Like the N cycle, the P cycle describes the dynamic nature of P movement within the environment (Figure 3.1). However, the P cycle, like potassium has no major gaseous component. Almost all P in terrestrial ecosystems is derived from weathering of minerals, such as apatite. Unlike N, the major source of mineral P is not provided by microbial reactions. In many ecosystems the organically associated P is the greatest source of available P for biogeochemical cycling (Schlesinger, 1997). While certain P solubilising mechanisms exists to increase availability in impoverished soils (Box 3.1) there is no process akin to N-fixation which can have such a marked affect upon N soil status. These factors coupled with the relatively low P content of most rocks and the limited bioavailability of P in soils has meant that many organisms in terrestrial and marine ecosystems survive due to the efficient recycling of organic forms of P.

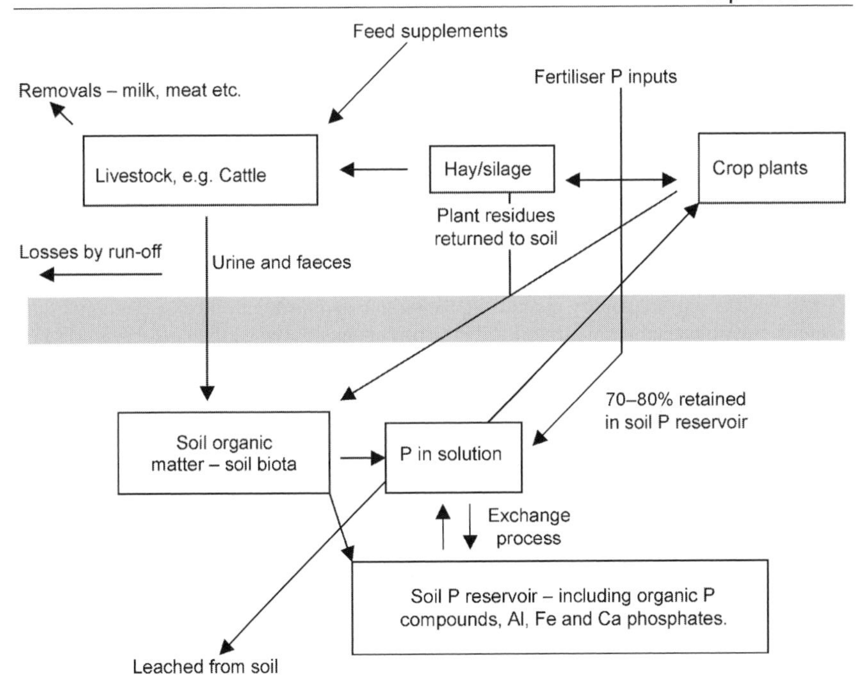

Figure 3.1 The agricultural P cycle (adapted from Glendinning, 1999).

Phosphorus inputs and outputs

Perturbations in the P cycle caused by anthropogenic activity are considerable, but the manifestations of these vary greatly in magnitude and scale, depending to a large extent upon the fluxes within the cycle. The management of soils for agricultural use has a considerable influence upon the P cycle, particularly with respect to the dramatic shift in the balance between P inputs and offtakes. Calculation of the annual agricultural P budget within the EU has shown that in all countries a surplus exists (P inputs to the farming systems such as inorganic and organic fertilisers are more than the rates of outputs in the form of crops and livestock). Belgium and the Netherlands have the greatest annual surpluses in the EU of 40 kg P ha^{-1} (Brouwer *et al.,* 1995). It has been estimated that in the EU the P status of agricultural soils has increased by 45 million tonnes during the last 30 years – equivalent to >10 kg ha^{-1} $year^{-1}$ (Tunney, 1992). In the UK, a national assessment of inputs and outputs of P to a range of farming systems suggested an annual surplus of approximately 16 kg P ha^{-1}.

One of the results of this imbalance between inputs and outputs (or 'offtakes') is a build up in soil P to levels that are of environmental rather than agronomic concern (Daniel *et al.*, 1998). The increase in soil P content over time was, and in some scientific circles, still is, thought to be beneficial to long-term soil fertility (Glendinning, 1999). In the UK, between 1979 and 1985, an increase in topsoil (0–15 cm) P of over 200 mg kg^{-1} has occurred under both arable and grassland systems, representing an increase of almost 20% (MAFF, 2000a).

Across Europe, grain production systems are regarded as having the greatest P surplus, with 269 kg ha^{-1}, compared with an average of all-farming systems of 19 kg ha^{-1} (Brouwer *et al.*, 1995). The importance of each farming category in terms of percent land area and percent P surplus is shown for the UK in Figure 3.2. In catchments where similar farming systems operate, such as pig and poultry production, the large P surpluses may be expected to present potential environmental problems to local drainage channels (Haygarth *et al.*, 1998; MAFF, 2000b).

The relationship between surplus and loss of P from agricultural land is not solely governed by the size of that surplus. Indeed, Edwards and Withers (1998) suggest that losses from farming systems are often independent of the surplus and often related to changes in farming systems which have reduced residence times within components of the P cycle and increased the rates of transfer between them. Such changes are thought to include increases in livestock density, area sown to winter cereals, area which is underdrained and a shift to slurry-based livestock systems, with expansion of indoor slurry based systems being a major culprit (Withers, 1996). Total and extractable or 'available' P loss is always greatest from cultivated compared to uncultivated soils.

This excess of P inputs over outputs in agricultural systems highlights the relatively poor efficiency of P usage within European agricultural systems and gives an indication of the potential for leakage. The following headings will address the individual components contributing to the P balance and assess their environmental importance and impact upon the P cycle.

Fertilisers

It has been broadly accepted in intensive agriculture systems that good crop yields require appreciable applications of P to most types of soil. The birth of the modern P fertiliser industry was some 160 years ago. The acidulation of rock phosphates with sulphuric acid to produce soluble phosphates in 1840 still forms the basis of the modern day process (Tisdale *et al.*, 1993).

Rock phosphates ($Ca_{10}(PO_4)_6(X)_2$, where X is either F, OH, or Cl) or apatites are the original source of P used in all P fertiliser manufacture. Although under some circumstances rock phosphates may be used directly for the supply of plant P if finely ground, for the most part they tend not to be used because of relatively low P content (14–17%), low solubility and inconvenience of handling compared with other P fertilisers.

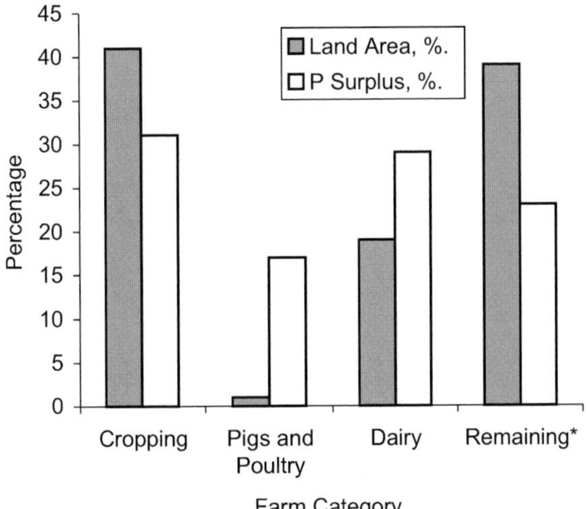

Figure 3.2 The contribution of farming types to the UK annual P surplus, as a percentage of the total agricultural land (Edwards and Withers, 1998). *Remaining includes mixed farms and horticulture.

Rock phosphate and the most commonly used agricultural fertilisers, their P content and availability are shown in Table 3.2. Other major elements contained in these fertilisers are also shown and may be used in the manufacture of 'compound' NPK fertilisers. Up until the early 1960s triple super phosphate (TSP) was the most commonly used P fertiliser source, especially in the USA, but since then the ammonium phosphates have become increasingly popular, although, in Australia and New Zealand, single super phosphate and TSP are still widely used. The popularity of ammonium phosphates is due not only to increased P content and solubility but also to the fact that increased plant P uptake occurs when ammonium is included in P fertiliser (Tisdale *et al.*, 1993).

The efficiency of applied P fertilisers in terms of crop uptake is dependent upon many soil, environment and management factors but is generally between 11 and 38% (Withers and Sharpley, 1995). While this may seem particularly inefficient, it does compare favourably with livestock production systems (Chapter 5).

It is important to note that many of the factors influencing native or indigenous soil P behaviour also affect fertiliser P additions. Therefore, whilst Table 3.2 gives figures for fertiliser P efficiency in terms of crop uptake these also vary upon addition to soil.

Atmospheric inputs

Atmospheric P cycling is, unlike N, limited, with no major gaseous phase. The annual cycling of P through the atmosphere as dust and sea spray is less than 1×10^{12} g, compared with 21×10^{12} g which is carried by rivers to the sea (Schlesinger, 1997). Rainfall tends to have very low P content (<0.1–0.05 kg ha^{-1} $year^{-1}$) and more P reaches the soil surface as dry deposition from wind blown dust or soil (White, 1997). In the UK, the annual atmospheric deposition of P onto soil varies according to geographic location, but is thought to range from 0.11 to 0.22 kg ha^{-1} for rural areas (Haygarth *et al.*, 1998).

Table 3.2 Common P containing fertilisers and their P content and availability (Tisdale *et al.*, 1993; Pierzynski *et al.*, 2000).

	Percentage of P content	Percentage of Total P available	Other Nutrients (%)
Inorganic P			
Rock Phosphate	14–17	14–65	variable
Triple Superphosphate	19–23	97–100	S(1–1.5%), Ca
Monoammonium Phosphate (MAP)	26	100	N(11–13), S(0–2)
Diammonium Phosphate (DAP)	23	100	N(18–21), S(0–2)
Organic P			
Dairy Manure	0.6	–	N,K,S,Ca, Mg and trace elements
Beef Manure	0.9	–	"
Poultry Manure	1.8	–	"
Pig Manure	1.5	–	"

Animal manures

In the UK, over 200 million tonnes of manure from livestock is produced annually, containing around 119,000 tonnes of P. The recycling of this manure to agricultural land (55% to arable land and 45% to grassland) is estimated to have a potential value of over £80 million based on the total P (Smith *et al.*, 1998).

Like N, almost 70% of the P consumed by cattle and over 80% consumed by sheep in concentrated feeds, fresh grass and silage is excreted (Haygarth *et al.*, 1998). From a range of livestock production systems in a selection of European countries the recycling of P was found to be between 10 and 34%. Phosphorus supplements in livestock feeds are particularly important, as too little P may cause reduced milk yields, poor live weight gains, reduced food intakes and fertility. However, too much will mean more P will be excreted, therefore a careful balance is required (Withers and Sharpley, 1995).

Phosphorus present in manure and slurry may be in a range of forms, from solid-phase inorganic P to dissolved organic P complexes. However, the composition and content of manure is highly variable and depends upon many factors including the proportion of mineral P in the animal feed, fertiliser P additions to grazed pasture and livestock type and age (Table

3.2). Storage of animal manures tends to increase the inorganic P content as microbial mineralisation breaks down the organic P fraction. While the P present in manures from livestock tends to be only slowly available compared with water-soluble inorganic fertilisers, it may be totally available in the long-term (Withers and Sharpley, 1995). This coupled with the relatively poor efficiency of P recycling in livestock suggests that large surpluses and potential soil accumulation may occur in areas where livestock production systems predominate.

P transformations in soil

Phosphorus transformations may involve the microbially mediated mineralisation and immobilisation reactions of soil organic P. Like N transformations in soil, the reactions associated with organically associated P are affected by a range of soil, organic substrate, climatic and seasonal factors (Sharpley and Halvorson, 1994). Both mineralisation and immobilisation reactions take place simultaneously, with the former resulting in the release inorganic P into solution and the latter the conversion of inorganic P to less readily available biochemical compounds. The net reaction depends upon total organic P and organic carbon contents of the residue. Some organic substrates are extremely resistant to microbial degradation, but almost all organic phosphates of plant origin may be broken down eventually by the phosphatase enzyme in microbes (Tisdale *et al.*, 1993).

Table 3.3 Effect of carbon and P ratios on microbially mediated organic P transformations (Soil Fertility and Fertilisers, 5/E by Tisdale, S. © Reprinted by permission of Person Education, Inc., Upper Saddle River, NJ).

C:P ratio	Mineralisation/immobilisation
>200	Net mineralisation and release of mineral P
200–300	Possibly no change in mineral P
<300	Net immobilisation of mineral P

Table 3.3 shows the importance of the ratio of the carbon to P content of the organic matter. The addition of organic manures and wastes to soils has been shown to elevate total inorganic P levels and not, as would be expected, total organic P, highlighting the role of mineralisation reactions.

Distribution of P in soil

The physical and chemical distribution of P within soils has an enormous bearing upon the environmental behaviour and subsequent management of P both in agricultural systems and the broader environment. Phosphorus in soils is present in inorganic or organic forms with total concentrations in topsoils of 100–3000 mg kg^{-1} (Laegreid *et al.*, 1999). Phosphorus may be categorised into a series of nutrient pools:

- non-labile inorganic P – particularly in Al, Fe and Ca minerals (or occluded P, only becoming available to plants and organisms very slowly);

- labile P – P adsorbed to soil particles;
- biomass P – P associated with living organisms;
- organic P – P associated with non-living soil organic matter;
- dissolved P – P in both inorganic and organic forms.

There are a range of organic P compounds in soils including nucleic acids, inositol phosphates and phospholipids, but only 50% of soil organic P compounds have been identified. Due to its close association with organic matter, organic P is found primarily in the topsoil and amounts decrease rapidly with depth. Apart from the benefit of being a potentially available source of P, organic P associations in soils are also considered beneficial by facilitating the plant uptake of P by forming organophosphate complexes which remain in solution and coating Fe and Al particles to reduce sorption (Tisdale *et al.,* 1993).

Although P is a relatively abundant element, its availability to plants and other organisms in soils may be relatively low. For example, in a relatively fertile soil the total P in the top 10 cm may be 480 mg kg^{-1}. However, the distribution of P into various soil pools is typically (mg kg^{-1});

Inorganic	stable	200
	labile	20
Organic	stable	200
	labile	20
Biomass	plants	20
	Micro-organisms	20
Soil solution		<0.1

Soil solution is considered the critical pool from which plants take up nutrients. Soil solution P concentrations are typically 0.05 mg l^{-1}, whereas concentrations required by plants are between 0.003 and 0.3 mg l^{-1}. Plants generally take up P as orthophosphate ions ($H_2PO_4^-$ or HPO_4^{2-} depending upon soil pH) from soil solution, but the soil solution pool of P can be a very small fraction of the total (<0.002%). Throughout a growing season crop plants will rapidly deplete this, hence the solution P must therefore be replaced on a regular basis as it is depleted (thought to occur at least 300 times during a growing season) to maintain sufficient P levels for plant growth. Figure 3.3 shows the inter-relationship between soil solution P, the 'labile pool' (i.e. the P adsorbed to surfaces that may be desorbed and the solution P) and the 'non-labile pool' which tends not to be available to plants.

This replenishment of P in soil solution is critical if soil fertility is to be maintained. However, while P in soil solution may be maintained and therefore be available for plant uptake, it may not be accessible to plants, therefore nutrient ions also move through the soil to the plant roots. Differences in water potential in soil gives rise to water movement in soils and ions or molecules present in solution will also move, the process is known as mass flow or convective flow. As there are relatively small concentrations of P in soil solutions, mass flow accounts for 1–20% of P transport to the root surface (Tisdale *et al.,* 1993). Mass flow is an important mechanism for NO_3^-, Ca^{2+} and Cl^- movement, but is limited in terms of P transfer to plant roots.

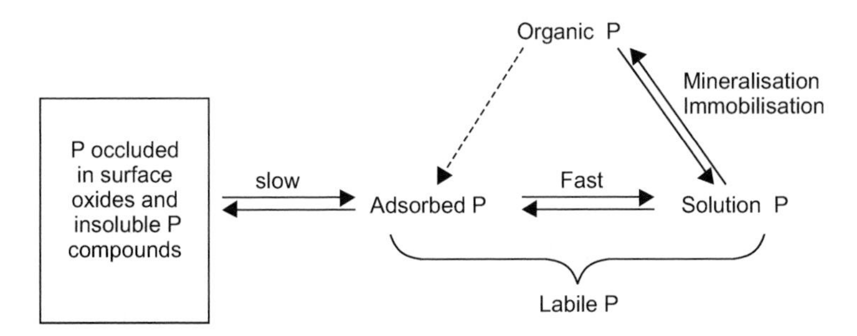

Figure 3.3 Transfers and transformations of P in soil, the arrows in the diagram give an indication of the direction P may move and the relative speed at which this may occur (Reproduced from White, 1997 with kind permission of Blackwell Science Ltd).

Consequently, diffusion, the movement of ions down concentration gradients, is the primary mode of P transport to the plant root. The diffusion coefficient is a term used to describe the movement of a solute from a region of high concentration (e.g. fertiliser pellet) to a region of lower concentration (e.g. in the vicinity of a plant root). The diffusion coefficient varies greatly between ions and with soil properties. Estimates of how far an ion will move through the soil in a period of 6 days are shown in Table 3.4 and highlight the limited distance orthophosphate moves by diffusion compared to the more mobile nitrate and potassium ions (Wild, 1981).

The relatively low mobility of P in soils may lead to the development of depletion zones around plant roots i.e. when the uptake of the nutrient by the plant exceeds the rate at which the labile pool can replace the solution P (Cooke, 1981).

Due to the complex relationships within soils that govern the distribution of P and ultimately the availability of P to plants, it has become essential to be able to estimate the amount of P in the labile or 'plant available' pool, as the total amount of P in a soil may have very little bearing on the fraction that is agronomically important. One of the main methods used to routinely determine the availability of P in soils from the field is by chemical extraction. By using a dilute chemical extractant, such as 0.5M $NaHCO_3$, (Olsen's reagent), the process of mineral dissolution that occurs when plants take up P from solution by precipitating solution Al and Ca is simulated (Tisdale *et al.,* 1993). The use of such extractants is undertaken following their calibration with plant tissue P analysis under various rates of P application. This calibration means the P solubilised by a particular extractant may be used to estimate P availability and fertiliser recommendations (Rowell, 1994) (Table 3.5).

The use of chemical extractants such as $NaHCO_3$ should only be considered as a guide only as site-specific factors often need to be taken into consideration if recommendations are being made. Some of the inherent flaws and potential problems associated with using and interpreting these tests are given elsewhere (Barber, 1995). For all of these problems, the use of chemical extractants gives a broad indication of an operationally defined measure of P availability

Table 3.4 Distance moved by macronutrient ions by diffusion in 6 days (in mm).

	Volumetric water content	
Ion	0.3	0.1
NO_3^-	30	3
K^+	3	0.3
$H_2PO_4^-$	0.3	0.03

Table 3.5 Fertiliser P recommendations (kg P_2O_5) for cereal crops in the UK using the Olsen or bicarbonate extractable P and availability index (MAFF, 1988).

Extracted P, mgl^{-1}, (and availability index)	0–9(0)	10–15(1)	16–25(2)	26–45(3)
kg ha^{-1}	0–23	24–38	39–63	64–113
For 6 t grain yield ha^{-1}	47	21*	21*	21*
For 10 t grain yield ha^{-1}	57	35*	35*	35*

*Application to maintain soil P status and replace crop removal.

P sorption

Phosphorus sorption is the term used to describe the adsorption/precipitation of P from solution as sparingly soluble solid forms. On addition of amendments to soil, P may be taken up by crop plants, held as organic P or become weakly or strongly bound with inorganic soil components. Precipitation reactions are responsible for P retention and removal from solution when the concentration of solution P and relevant cations exceeds the solubility product of the mineral. Adsorption is more likely to account for P removal from solution when P solution concentrations are low, although, it has been suggested that P retention is probably a continuing mix of both mechanisms (Tisdale *et al.,* 1993; Rowell, 1994).

The sorption of P from solution is initially rapid, followed by a slower conversion to progressively less plant available forms of P, including insoluble Fe, Al and Ca phosphate minerals, depending on soil pH (Figure 3.4).

Sorption of phosphorus is influenced by a range of soil physico-chemical factors including soil texture and mineralogy, soil pH, cation and anion effects, organic matter and reaction times and temperature (Tisdale *et al.,* 1993). The retention of P in soils has important implications for the residual value of P fertilisers, where 70% of that added may be fixed in a single growing season. A relatively small proportion of this fixed P may be available for plant uptake in the next growing season. Studies in New South Wales in Australia have shown this may be as little as 24% (Glendinning, 1999). Whereas others have observed this figure to be somewhere between 6 and 70% depending on the type and form of P fertiliser added and land management factors (Larsen, 1971; Sharpley and Halvorson, 1994). In some soils with a high P sorption capacity, large-one off P applications (>200 kg ha^{-1}) (Table 3.5) followed by normal annual rates are required to increase the soil solution P concentration above the critical limit for crop growth (Laegreid *et al.*, 1999).

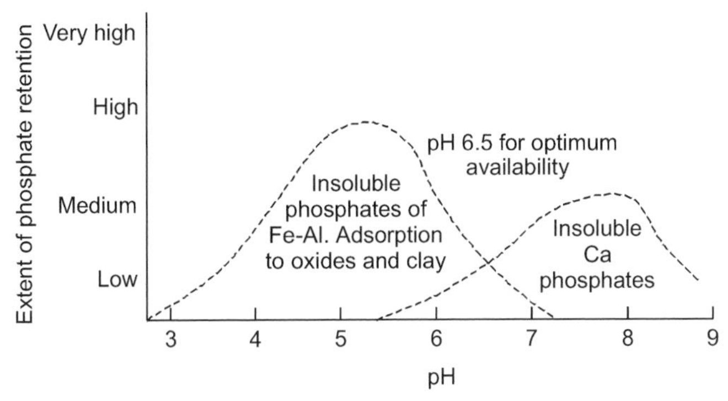

Figure 3.4 The influence of pH on the retention of P (Soil Fertility and Fertilisers, 5/E by Tisdale, S. © Reprinted by permission of Person Education, Inc., Upper Saddle River, NJ).

The retention of P by sorption can influence solution P concentrations so will also influence plant uptake and subsequent fertiliser management, affecting the movement and loss of P from the agroecosystem.

Box 3.1 Utilising indigenous soil P

Due to the inherently low mobility of P in soil, some organisms have developed mechanisms to improve supply. These range from the secretion of organic acids and solubilising chemicals to symbiotic fungal associations. For example, lupins (*Lupinus albus*) secrete citric acid that precipitates calcium, promoting the release of soluble P from insoluble phosphates (Laegried *et al.*, 1999).

Mycorrhizas are symbiotic associations between a fungus and the plant root. The relationship is mutually beneficial with the fungus growing from the plant root and taking up increased P (and also Cu, Zn and Co) and supplying it to the plant in return for carbohydrate (Rowell, 1994). The mycorrhizal infection increases the soil volume exploited by the roots (in some cases through an extensive network of hyphae amounting to 1–20 m^{-2} g^{-1} of soil) and improves P storage and release in times of deficiency (White, 1997). Over 80% of all plant species may form mycorrhizal associations, however these tend to work more effectively in P deficient soils. Industrial usage and field scale manipulation of mycorrhizal inoculants have shown only limited success.

The commercial development of plant-beneficial P solubilising fungi is not limited to mycorrhiza, 'Provide', a biological product sold in Canada contains *Penicillium bilaii* which solubilises soil bound P (Cunningham and Kuiack, 1992). In Australia, a similar fungus *P. radicum* (sp. nov.) has been observed to increase wheat yields by 14% in the field and laboratory by stimulating plant growth directly via improved P nutrition (Whitelaw *et al.,* 1997). The success of these under European conditions is yet to be determined.

3.3 PHOSPHORUS LOSSES

There are a number of pathways by which P is lost from an agricultural soil, with the most desirable route being via crop uptake and subsequent removal by grazing or harvest. It follows from the above discussion of P behaviour within soil systems that little if any P would be transferred from soils or 'leak' into adjoining ecosystems. Consequently, the loss of P from agricultural systems is relatively low, certainly in comparison to the agronomic requirements for P. However, concentrations of P as low as $20\,\mu g\,l^{-1}$ can cause environmental problems in some waters, it is clear that even limited movement of P, be it soluble or adsorbed onto soil particles or organic matter, may be of considerable importance (Figure 3.5).

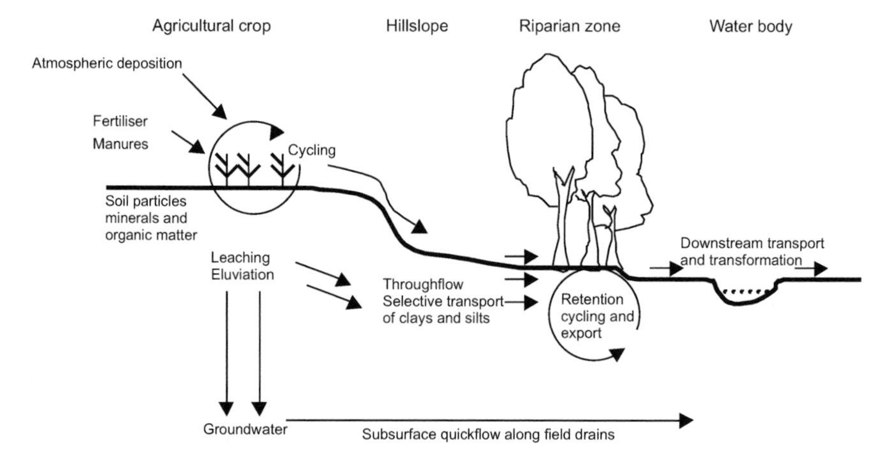

Figure 3.5 P inputs from non-point sources and main hydrological flow pathways (Johnes and Hodgkinson, 1998).

There are a number of point sources of P contamination of the environment (Box 3.2). The loss of P from agricultural systems may occur from point sources including accidental spillage from manure stores, livestock housing, direct contamination during spreading of manure or inorganic fertilisers and consented discharges from livestock farms (Withers and Sharpley, 1995). Point sources are easily identified and have therefore seen greater and more effective controls placed upon them compared to diffuse or non-point sources of P (Sharpley and Halvorsen, 1994; Daniel, *et al.*, 1998). The reduction in losses from these point sources of P has drawn greater attention to the non-point source forms of P loss from agricultural systems – i.e. losses from agricultural soils. The P may be transported from soils by a range of mechanisms including surface runoff, soil erosion, leaching and subsurface flow (Figure 3.5).

Phosphorus is transported from agricultural soils as either dissolved or particulate forms and generally a mix of both, depending on a range of soil, hydrology and land-use factors (Table 3.6).

Surface run-off

Surface run-off is the main mechanism of P loss from agricultural systems, and may account for 90% of the P transported from arable land in the UK, although considerable problems arise in quantifying the magnitude of diffuse inputs (Catt *et al.*, 1998). Phosphorus losses can be from a number of source areas in an agricultural catchment. The form, quantity and temporal variability of losses are influenced by a host of short-term changes in farm practice and hydrological conditions (Sharpley *et al.*, 2000).

The loss of P via surface run-off is influenced to a large extent by the rate, timing, form (inorganic manure, type of manure etc.) and method of P fertiliser applications, as well as antecedent and post application rainfall and farm type (Sharpley and Halvorsen, 1994). In addition to these factors, catchment properties that control soil, stream bed and channel bank erosion and sediment transport such as steepness of slope, surface roughness and other soil physical properties (aggregate stability, dispersion, texture, etc.) will also have an effect on P loss.

Water flowing across the surface of the soil may entrain particles and desorb and transport soluble P. The majority of P in run-off is in the particulate form, especially if the run-off is dominated by suspended sediment – which tends to occur during storm events (Pierzynski *et al.*, 2000). The association of P with fine particulate material was observed by Edwards (1996) in a cultivated soil which had received regular inorganic P applications, where the <0.1 mm particle size fraction had 1.5 times more total P compared with the 1–2 mm fraction. However, the ratio of 'total P: available P' declined from 49 to only 14.

This reduction in available P occurs due to the increased sorption on the relative large surface areas of the fine material (i.e. clays sized <2 μm) being transported. It also explains the relative enrichment of P in eroded particulate material compared to the source soils due to the preferential transport of fine particles (Withers and Sharpley, 1995). Because of this process the enrichment ratio is commonly used to describe the relative difference between the concentration of P in sediment removed from soil as rainfall run-off and the concentration of P in the source soil. Figure 3.6 shows the enrichment ratios for P in rainfall runoff for two soils (1 and 2) from the USA treated with a range of inorganic P applications and subjected to simulated rainfall. An increase in the fertiliser application rate increased the enrichment ratio and the P content of the run-off sediment compared to the source soil. The increase in the enrichment ratio even in the zero treatments, highlights the preferential movement of fine particulate material, commonly associated with greater total P content (Sharpley, 1980).

Table 3.6 Factors responsible for controlling P loss by erosion and leaching on a range of scales (Edwards and Withers, 1998).

Scale	Leaching	Erosion	Leaching and erosion
Soil Profile	Soil sorption properties, pH, soil solution ratio	Aggregate stability, soil texture	
Field	Soil mineralogy, P availability index, artificial drainage	Extent and nature of crop cover, cultivation practice	Rainfall intensity, antecedent moisture content
Farm	Farm type, P surplus	Proximity to river, slope, field boundary conditions	Soil type, farm type
Catchment			Climate, land use, soil type, relief, farm types and numbers

Soluble P makes up a greater component of P in run-off from sites that frequently receive applications of organic manures and slurries. Up to 20% of the total P in poultry manures may be water soluble and contain a considerable quantity of low density organic material which greatly increases the susceptibility to water entrainment. The pattern of P distribution in run-off changes if the manure is deposited onto land directly from livestock rather than being spread, with greater total and particulate P losses tending to occur when stocking densities are high (Withers and Sharpley, 1995).

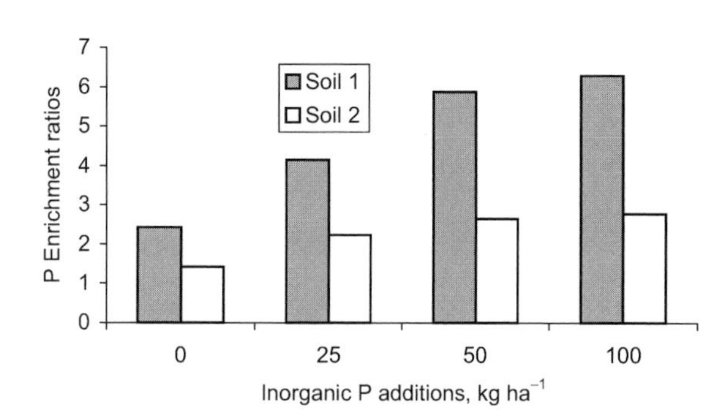

Figure 3.6 The enrichment ratios of sediment run-off as a function of fertiliser P application (Sharpley, 1980).

The temporal variation in P losses via run-off from soils receiving manure may be considerable. Averaged over a year, losses of P from agricultural land are unlikely to produce a situation where detrimental environmental effects occur. However, single run-off events following manure or slurry applications can have total P concentrations of up to 25 mg l^{-1} (Smith *et al.*, 1998). The critical concentration of total P entering water bodies to prevent accelerated eutrophication has been set by a range of organisation at 0.1 mg l^{-1} (Withers and Sharpley, 1995). Losses of P from manures and slurries via surface run-off are also promoted by a capping or sealing effect that can occur on land receiving high slurry applications (>80 m^3 ha^{-1}). This effectively reduces rainfall infiltration, heightening the risk of run-off. The greatest likelihood of P loss from farming systems occurs when (Withers, 1996):

- the catchment has erodible soils;
- the levels of P in the soils have accumulated to excess;
- rainfall closely follows fertiliser or manure application.

Phosphorus surplus in topsoils have increased the potential for loss of P from agricultural systems. These loses reduce the efficiency of P utilisation in terms of providing an agronomic return and provide evidence that calls into question the sustainability of management practices that accelerate the loss of a finite resource in a non-recoverable and environmentally degrading form.

Leaching and drain flow

Solution concentrations of P in most soils are extremely low and the distances moved by P ions in solution are so small that leaching to ground and surface waters would be expected to take many thousands of years. Unlike NO_3^-, the leaching of P is rarely thought of as being an environmental issue (Pierzynski *et al.*, 2000).

There are exceptions to this general observation. Whilst some are related to soil types (organic soils with fluctuating water tables and soils with well-developed macropore structure) pH and weather conditions, it has been observed that fertiliser usage may also be important in P leaching from agricultural land. In experiments performed in the Broadbalk Continuous Wheat Experiment at Rothamsted, Heckrath *et al.* (1995) suggested that leaching and subsurface runoff from fertilised soil plots could be greater than previously thought and may be related to soil P surplus. Samples of drain flow from pipes under the middle of fields receiving a range of inorganic fertiliser and manure amendments were collected for two years. Concentrations of total P in the waters draining through the soils ranged from 0.03 to 0.23 mg l^{-1} in unfertilised plots and up to 0.55–2.75 mg l^{-1} in inorganically fertilised plots. The highest values of $NaHCO_3$ extractable P (90 mg kg^{-1}) were also observed in the soils receiving inorganic fertiliser treatments, and a strong positive correlation was observed between values of $NaHCO_3$ extractable P (when >60 mg kg^{-1}) in the plough layer and P in the drainage water. This increase of P in drainage waters is thought to reflect the long-term accumulation of P in soils, beyond the level likely to result in beneficial agronomic effects (Sharpley *et al.*, 2000).

The use of artificial drainage in heavy or organic soils increases infiltration and decreasing the potential for runoff, but may increase the chance of leaching. Compared with natural subsurface movement, drains greatly accelerate P transfer. Considerable losses of P may occur via subsurface pathways, including field drains (>2.5 kg P ha^{-1} $year^{-1}$) in arable land, which would be sufficient to cause water quality issues (Catt *et al.,* 1998). It has been estimated that almost 50% of the productive agricultural land in England and Wales is underdrained, with the primary reason being the reduction of soil water and allowing autumn access to sow winter cereals (Withers *et al.,* 2000). Unlike surface run-off which may be intercepted before reaching surface drainage channels by landscape features, such as hedgerows or riparian buffer strips, a field drain will usually lead directly into an open drainage channel.

Large applications of animal manures to coarse sandy textured soils (these soils often have very low sorption capacities for P), necessitated by shortages of land for appropriate disposal, may result in increased concentrations of P migrating below the plant rooting zone to ground waters and field drains. Increased P movement via leaching has also been observed when slurries are injected into grassland soils (Tunney *et al.,* 1997). Subsurface drains and preferential flow pathways such as macropores (Brookes *et al.,* 1997; Hooda *et al.,* 1999) act as conduits for the transport of P which by-pass the potential sorption sites in the subsoil that would otherwise reduce P movement and transport (Withers and Sharpley, 1995; Johnes and Hodgkinson, 1998). Consequently, the result may be an increased transfer of P, at concentrations of up to 10 mg l^{-1} in a single storm event to local waterways (Smith *et al.,* 1998).

Leaching and subsurface loss of P from agricultural systems would suggest that P is being transferred primarily in the dissolved form. This is not always the case; finer soil fractions may also be transferred, with P sorbed to soil colloids, along subsurface pathways. The proportion of P transferred in particulate form can range from 17 to 60% of the P lost via drainage water (Heckrath *et al.,* 1995; Hooda *et al.*, 1996) and is dependent on a range of soil, management and hydrological factors. This is further complicated as during the transfer of P from soil to water, the form of P may continuously change between the dissolved and particulate phases.

3.4 ENVIRONMENTAL IMPLICATIONS OF P POLLUTION

We have demonstrated that post 1945 agricultural intensification in many European farming systems has lead to an increase in soil P content to levels beyond those that would realistically benefit crop and grassland production (Johnes and Hodgkinson, 1998). This combined with a range of other factors has resulted in significant proportion of the P applied to agricultural land in the form of manures and inorganic fertilisers being transferred via a variety of routes to surface and ground waters. The environmental implications of this transfer of P and resulting enrichment will now be considered.

N and P balance in natural waters

It is important to note that water quality issues, such as eutrophication, are not solely caused excessive P inputs. A host of environmental factors including water salinity, temperature, depth and light as well as the presence and interaction of N and P are all involved (Pierzynski *et al.,* 2000).

The influence of N upon primary biological productivity in aquatic systems, particularly coastal waters and estuaries, was discussed in Chapter 2. However, in freshwaters the biologically limiting element is almost always P. The importance of the balance between N and P in terms of primary production, such as algal growth, in aquatic systems has long been recognised. Under favourable growth conditions algae have an N and P elemental composition that falls within a set ratio. This is known as the Redfield ratio, and commonly falls between 15 and 16:1 (Correll, 1998). However, it does vary with a range of environmental conditions, when values exceed 16:1 it has been suggested that P will be the element limiting algal growth. Conditions which tend to favour water quality problems are when the total P concentration of the water body is >0.05 mg l^{-1} and the Redfield ratio of N:P is 20 or less (Peirzynski *et al.*, 2000). Aquatic systems naturally low in P are very sensitive to inputs of P.

The inter-linkage of P and N in freshwater algal populations demonstrates the importance of an holistic fertiliser management strategy on a range of scales if nutrient leakage from agroecosystems is to be reduced and problems with water quality are to be tackled successfully.

Box 3.2 Point sources of P loss

Phosphorus may enter aquatic systems from a range of point sources including urban waste water drains (including P containing detergents and sewage wastes), sewage-treatment works, fish farms, meat, dairy and food processing plants (Lennox *et al.,* 1997). In the EU, it has been suggested that over 50% of the anthropogenic P entering aquatic systems is from urban waste-water (Mariën, 1997), with a strong correlation between population density and P concentrations in river waters. The highest concentrations (often 80% of the rivers have >0.125 mg P l^{-1}) in European rivers are in Northern Ireland, southern England and across central Europe through Romania to the Ukraine (Steén, 1997). Agricultural non-point source losses of P account for less than 50% of the total (of which 16% is thought to originate from fertilisers and 34% from livestock), but this is the hardest and most costly to control (Environment Agency, 2000a). It is estimated almost 90% of the P entering rivers in East Anglia is from sewage works effluent. Areas which have been particularly susceptible P enrichment from this source also include Loch Leven in Scotland, Lough Neagh in Northern Ireland, the Shropshire and Cheshire meres and a number of water supply reservoirs in southern England (Foy and Bailey-Watts, 1998). In 1999, 23% of all substantiated water pollution incidents (those in which a prosecution followed) in England and Wales were caused by sewage, only those caused by fuels and oil were greater at 26% (Environment Agency, 2000a). However, the inputs of P from point sources are easier to identify and so have seen more attempts at control than diffuse or non-point sources as industrial and sewage discharges are brought into line under the Urban Waste Water Treatment Directive (Heathwaite, 1997; Foy and Bailey-Watts, 1998). Yet there is limited evidence to suggest that the reductions of P input from these point sources has had any dramatic effect on P content in aquatic systems, suggesting diffuse sources of P (or even P release from sediments) may be playing a critical role (Steén, 1997).

3.5 ENVIRONMENTAL PROBLEMS CAUSED BY P POLLUTION

The overriding concern associated with P loss from agricultural land is the pollution of surface waters causing eutrophication. However, a further issue is the loss of indigenous vegetation types in naturally nutrient poor grassland systems which have received increased P inputs. The human health effects of excessive P are, unlike N, primarily limited to those effects resulting from impacts on surface water quality.

P in ground and surface waters

Phosphorus concentration in ground waters tends to be relatively low and subsequently of limited environmental importance (Sharpley *et al.,* 1987). Nevertheless, the total concentration of P in surface waters (often considered critical in terms of the potential to accelerate biological productivity and potentially eutrophication) is only 0.01 mg l^{-1} (cf. N 1–0.5 mg l^{-1}) (Daniel *et al.,* 1998). This is 10 times lower than the concentration in soils commonly thought to be required for crop growth. This difference is of fundamental importance in regard to issues of water quality and the management of P usage in agricultural systems. The loss of P from these systems is of limited agronomic value, but crucial in regard to environmental quality.

The effects of these losses depend upon the characteristics of the receiving water. For example in rivers, where ecology is largely determined by flow regime, residence times may be too short for phytoplankton to develop. Favourable conditions may be reached when the speed of the water is checked, such as in reservoirs, dams or slower moving reaches (Gibson, 1997). The biological responses in streams and rivers to increased P loading are site and season specific (Edwards *et al.,* 2000). In the UK, rivers flowing through areas dominated by arable farming tend to have greater P concentrations than similar lowland areas in which pastoral activity is the main use (Environment Agency, 2000a).

Increases in P in surface waters may cause rapid rises in the growth of algae and other water plants. Indeed, a linear relationship is thought to exist between biomass production and concentrations of total P in waters of less than 0.1 mg l^{-1}, but above this other factors such as light availability become more important (Gibson, 1997).

A schematic of the relationship between primary production, the input of P and the trophic status of a freshwater system is given in Figure 3.7. Generally, there are four consecutive stages of eutrophication; oligotrophic, mesotrophic, eutrophic and hypereutrophic. The progression through these stages results in changes in the biodiversity, ecology and oxygen status of the water body (Peirszynski *et al.,* 2000).

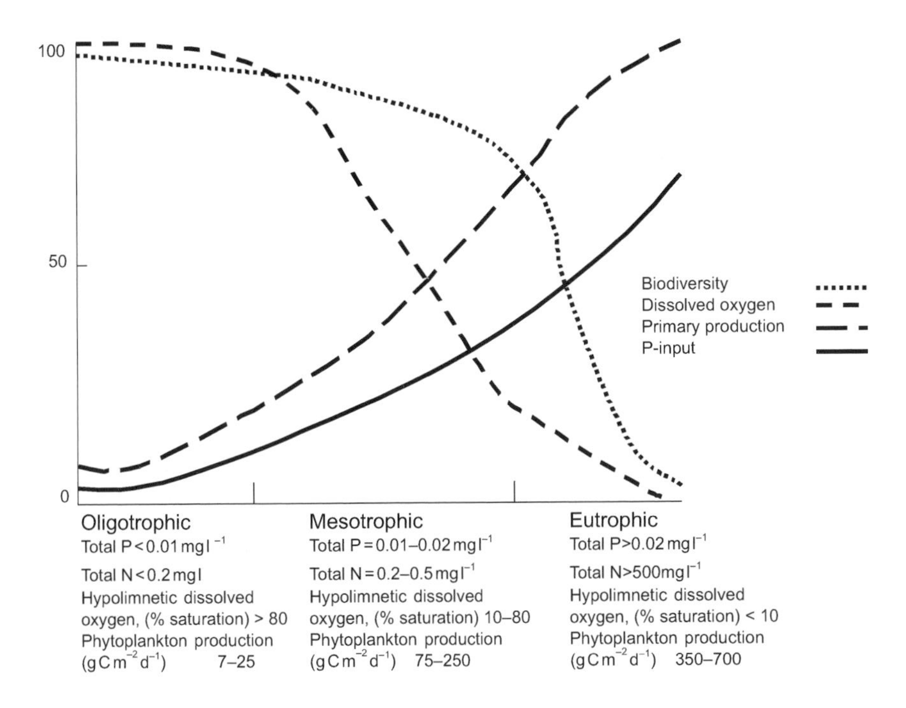

Figure 3.7 The relationship between P input and biodiversity in freshwater systems, the values below each trophic status are examples of possible water quality characteristics (adapted from Correll, 1998 and Mason, 1996).

On reaching surface waters the form of P can change, so while dissolved P may be the form in which it reaches a lake, it may well be sorbed onto sediments and organic matter in the water column (Figure 3.8). Soluble P may represent an immediately bioavailable source on reaching a receiving water body. Particulate P is a long-term source to the aquatic system and the rate of P release is determined by the physico-chemical characteristics of the specific sediment and water column (Sharpley *et al.,* 1987). In nutrient poor aquatic systems, for example a freshwater lake, the P will remain predominantly in the bottom sediments. However, if this system receives a large input of P it may become eutrophic. The resulting increase in primary production can promote anoxic conditions in the water close to the bottom which causes the increased release of P from the sediments back into the water column (Correll, 1998). A common pattern of P cycling in lakes is that in winter most P is in the dissolved form, but as spring begins the increased light favours photosynthetic activity and phytoplankton populations increase and dissolved P concentrations fall (Gibson, 1997).

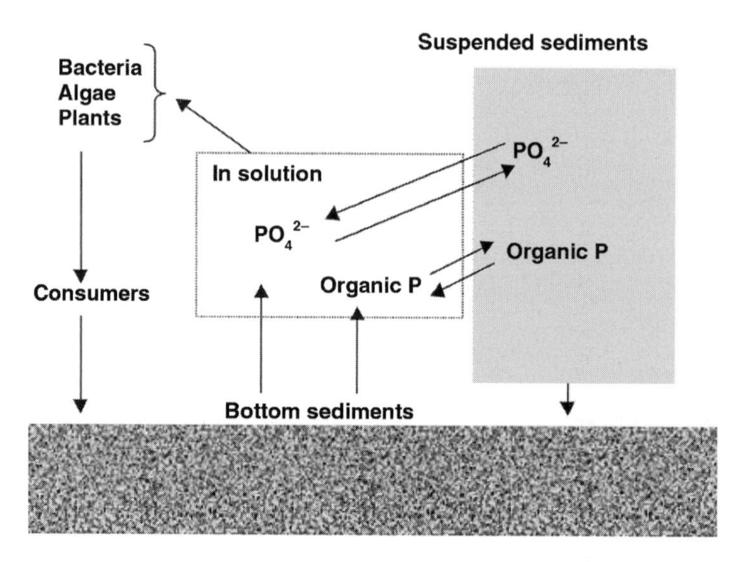

Figure 3.8 The potential cycling of P between various forms in aquatic systems (adapted from Correll, 1998).

The P transformations occurring in feeder channels may impact greatly upon the biological productivity of the receiving lake, reservoir or stream. This is very important, as net P input to a lake or stream from agricultural non-point sources may increase the chances of water quality problems such as eutrophication, but the transformations of P between the various forms affecting bioavailabity will determine the extent to which nutrient enrichment will occur.

Eutrophication

The eutrophication of surface waters is a global issue and is a result, in part, of elevated P concentrations. It can be defined as 'an increase in the nutrient status of natural waters that causes an accelerated growth of algae or water plants' (Pierzynski *et al.*, 2000). It is essentially a natural process, but over stimulation through anthropogenic inputs (cultural eutrophication) results in an increase in algal growth and plant production (Figure 3.7). It is common for lakes to have a peak in the population of diatoms (*Bacillariophyta*) in spring with a late spring flush of green algae (*Chlorophyta*). However, in eutrophic systems one may expect a large summer peak in blue-green algae (Cyanophyta) (Mason, 1996).

Approximately 6% of river lengths in the UK are designated as Eutrophic Sensitive Areas and instances of blue-green algal blooms are reported from over 200 sites a year (Environment Agency, 2000a,b). The long-term change in the trophic status of freshwater systems in the UK is difficult to determine, but between 1994 and 1998 a doubling in the number of designated sites sensitive to eutrophication has occurred.

Accelerated eutrophication of aquatic systems is associated with surface rather than subsurface inputs of P, a consideration of importance in the following discussion of management objectives to reduce P leakage. The detrimental effects associated with eutrophication of surface waters in the UK were outlined in Chapter 2 and include restrictions in water use for fisheries, recreation, drinking for humans and livestock and industry. Further discussion of these issues are given in other texts (Harper, 1992; Sakamoto, 1996; Moss, 1998).

3.6 PRACTICAL SOLUTIONS

The optimisation and effective use of soil P surpluses and fertiliser inputs are required for the sustainable use of P in productive agricultural systems. An improvement in understanding in these areas will enable the development of measures to reduce the loss of P from agricultural land and the subsequent transfer to surface waters (Sharpley *et al.*, 2000).

As the primary mechanism of P transport is via run-off, both dissolved and particulate P forms, the reduction of non-point sources of P requires management strategies that address both the sources and transfer mechanisms. Source management, such as better farm-scale nutrient budgeting and balancing P inputs for livestock and crop production may offer considerable financial savings along with environmental benefits. Transport management focuses on the reduction of the main P transfer pathways through the use of soil conservation, land-use management and buffer zones (MAFF, 1998).

It is clear from the preceeding chapter on N, that many of the management practices set out as part of the Code of Good Agricultural Practice for the Protection of Water (MAFF, 1998) will also apply to the reduction of P leakage. However, the Code also recommends that for soils containing over 25 mg kg^{-1} Olsen P the rates of manure application match P crop offtake and that no P is applied to soils containing >45 mg kg^{-1} Olsen P (soil P Index 3 – Table 3.5). Furthermore, nutrient budgeting targets are given along with guidance for the environmental suitability of changes in management on soils with a high soil P index (Withers *et al.*, 2000).

Concern in regard to eutrophication and the impact of non-point source P losses from agricultural systems at the national level is acknowledged by the publication of a national strategy – Aquatic Eutrophication in England and Wales: a Management Strategy (Environment Agency, 2000a). This document outlines an holistic approach to aquatic eutrophication, which requires action under a range of statutory and international commitments entered into by the UK (EC directives, UK Biodiversity Plan, the OSPAR Convention). The Environment Agency strategy is to focus primarily on N and P and their impacts upon ‘controlled waters’ (lakes, streams, canals, reservoirs, estuaries and coastal waters), and the overall objectives are to:

- provide a framework to manage cultural eutrophication, to protect, rehabilitate and restore waters adversely affected;
- deliver on the Environment Agency's 'Environmental Vision', particularly its commitments to eutrophication control;
- ensure contribution to the UK Biodiversity Plan, especially implementation of Habitat Action Plans for standing waters;
- promote a reduction in nutrient content of waters nationally, coordinate catchment based management of policies and procedures for the assessment and control of local eutrophication.

The document outlines a series of actions that will be performed to achieve these objectives which include:

- the setting up of eutrophication control action plans through the Environment Agency's catchment-based local Environment Agency Plan;
- promote the use of recycled phosphorus from soaps and detergents;
- refine and promote the uptake of the current codes of practice;
- with other organisations (e.g. LEAF) provide demonstration farms of good nutrient management and commercial success;
- review current eutrophication monitoring procedures via the National Collaborative Forum on Environmental Monitoring;
- further work on assessing the cost-effectiveness of current eutrophication control;
- provide interim targets (annual means) for P in freshwaters (eutrophic in standing water – 0.085 mg P l^{-1}, running water – 0.200 mg soluble P l^{-1}).

A range of practical strategies and options for reducing P leakage from agricultural systems currently exist and several examples are discussed below and shown in Figure 3.9 (Withers *et al.*, 2000). At field level, costs are generally high, but the methods effective. However on a catchment scale losses can be greatly reduced at relatively low costs. Increasingly approaches focus on an integrated catchment management plan, whereby site and catchment specific factors are taken into account and risk areas targeted.

Source management – optimisation of fertiliser P use

The reduction of P surpluses in soils is a long-term issue and rectification will take many years. Nevertheless, targeted use of fertilisers and nutrient management planning (and budgeting) may be effective in the short-term in reducing P losses through run-off (Sharpley *et al.*, 1993). The key aspects of P fertiliser management include:

- accurate estimates of crop and forage plant P requirements;
- use of soil, manure and plant tissue testing;
- monitoring of soil P levels;
- timely and efficient fertiliser practices (Daniel *et al.*, 1997; Pierzynski *et al.*, 2000).

Many soils in the UK have a P surplus and the greatest risks of further accumulation are where there is a limited land area for manure disposal. In particular, this occurs where manure applications are limited solely by N requirement – the ratio N:P is often lower in manure (e.g. 3) than is removed in the crop (e.g. 8) and so N tends to taken up, where as P may build up in the soil (Sharpley *et al.*, 1993; Smith *et al.*, 1998). It is therefore necessary that farmers adequately account for the P contribution in manures (even basing manure applications on P requirement not N) with the framework of an appropriate fertiliser plan (MAFF, 2000b).

Following the determination of a P requirement then the rate, timing, method and amount of fertiliser application need to be determined. The form of P fertiliser is thought to have a limited impact on P loss via run-off if the fertiliser is incorporated to the soil (Pierzynski *et al.*, 2000). Inorganic fertiliser applications may be placed and banded with crops to ensure more efficient usage in terms of crop uptake and crop recovery of applied P and less chance of loss through run-off (Tisdale *et al.*, 1993; Sharpley and Halvorsen, 1994). Estimates of P loss in runoff suggest that fields receiving broadcast P may be two orders of magnitude greater than from fields in which the fertiliser has been incorporated into the soil. Similar results have also been observed for broadcast and incorporated manure applications (Sharpley *et al.*, 1993).

Greater challenges exist in regard to manure usage and P loss. Best management practice for the recycling of animal manures and organic wastes to land are outlined in Chapter 5. However, specific measures to reduce P loss include composting, pelletising, transport from P surplus to deficient areas and the use of additions such as alum which can reduce P solubility (Sharpley *et al.*, 2000). Improved utilisation of the residual P content in soils and the more effective use of fertiliser applications could be possible through crop selection/rotation tillage and plant stubble management.

Animal feeds

Livestock feed is often seen as one of the major contributary factors to excessive P input into soils. Like the reduction and targeted use of P in fertilisers, the reduction of P inputs into livestock feeds is unlikely to have an impact in the short-term, as soil concentrations will change only slowly (Withers *et al.*, 2000). Therefore, this may be considered as a long-term preventative measure in comparison to the soil conservation and transfer management strategies below, which are likely to have immediate measurable impacts on leakage rates.

The manipulation of dietary P intake of livestock includes methods for increasing P adsorption by the animal, and refining animal feed rations. This issue is clouded by the variability in P availability in forages, feeds and supplements and the clear need to balance production and animal performance. Purchased feeds account for about 60% of the P inputs into intensive dairy operations and are mostly consumed as phytate. It is thought that limiting the amount of P fed to livestock is only going to be effective in reducing loss in intensive farming systems. Reducing stocking densities, the use of inorganic P fertiliser applications and using less purchased feeds would probably have a larger effect on reducing the P surplus and subsequent losses (Valk *et al.*, 2000).

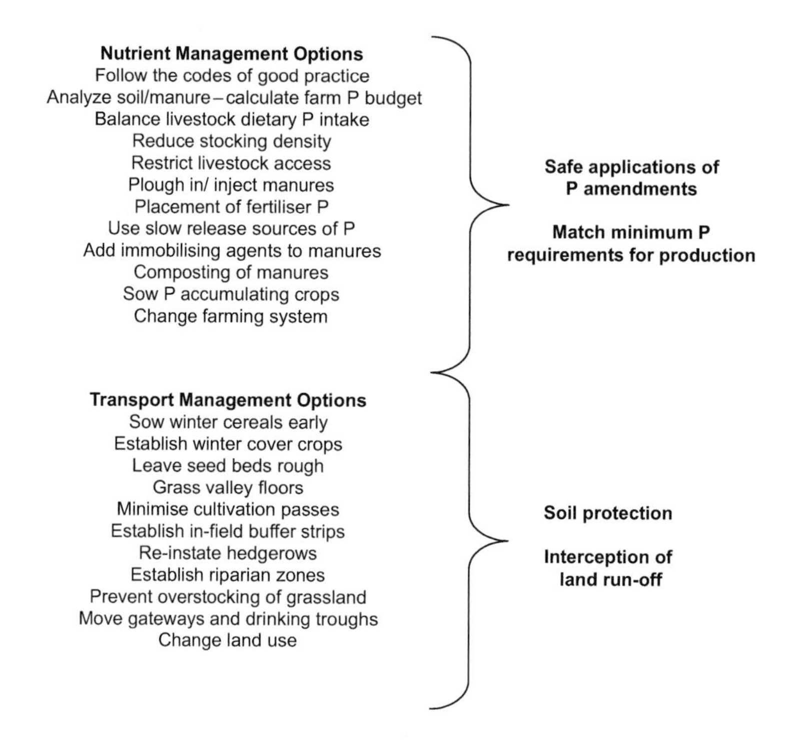

Figure 3.9 The nutrient and transport management options for the control of P loss from non-point sources (adapted from Withers *et al.*, 2000).

Nevertheless, the amount of P in livestock feeds need to be balanced with dietary requirements. Recent studies in the Netherlands suggested that feeds for dairy cows contained up to 20% more P than was required, due to the P content of grass silage. An overall reduction of dietary P to dairy cattle of 10% could be made (20% in the UK) without impacting upon production (Valk *et al.*, 2000). A further strategy that is currently being adopted to increase dietary utilisation of P by pigs and poultry is the supplementation of grain feed with phytase enzymes. This enzyme is absent in the guts of these animals and enables the digestion of a greater proportion of the P from the grain (the phytate P). Furthermore, grain varieties, low in phytic acid, are now being grown as feeds to reduce the need for the enzyme addition. Both these measures may reduce P in excreta by over 20% (Pierszynski *et al.*, 2000).

In the Netherlands there is combination of approaches being adopted to reduce excessive use of P. The policy is to bring about a stepwise reduction of the P supply towards an equilibrium level based on calculated removals and known inputs (Van der Molen *et al.*, 1997). The equilibrium level is defined as the sum of the manure and fertiliser supply that meets the need of the crops and compensates for inevitable losses. This approach defines losses under 'good agricultural practice' which are often higher than those losses considered acceptable in order to meet environmental standards. It is estimated that a relatively small excess of P inputs, in the order of 5 kg P ha^{-1} $year^{-1}$, will increase the area of land in the Netherlands that is defined as strongly P-saturated by 100% by the year 2050. In some parts of the country where P-induced eutrophication is a severe risk, manure-P returns to land are restricted based on a manure quota (Hotsma, 1997), with excess being exported to other regions with a lower P input. This is just part of a strategy being employed in the Netherlands where the intensity of livestock production has lead to serious environmental problems. Other approaches include the restoration of buffer zones and wetland margins, lowering groundwater levels and applying ferric and aluminium compounds to soils (Van der Molen *et al.*, 1997).

P transfer management by soil conservation

Soil conservation practices promote environmental and economic sustainability, irrespective of the potential influences upon P losses. In a recent study of 13 erosion susceptible catchments in the UK (MAFF, 1997) erosion occurred in 40%, of which the majority where cropped to winter cereals. The main factors associated with this erosion were poor crop cover (<15%), features that concentrated run-off and compacted tramlines (Chambers *et al.*, 2000). Soil erosion processes and control are discussed in more detail in Chapter 4.

Soil conservation techniques have commonly been focussed upon minimum tillage systems, well-timed cultivations and creating rough seedbeds. If these correspond to farming systems from which P is lost primarily in the form of run-off, then reductions in transfer should occur. However, results from the use of conservation and minimum tillage have been found to be mixed. For example soil erosion may be reduced, but increases in P loss through increased soluble P concentrations can occur (Gaynor and Findlay, 1995).

Whatever technique is adopted, it must be cost-effective. An enormous range of techniques can and have been adopted to reduce phosphorus losses in run-off and have been made viable through a range of schemes in England and Wales, such as Environmentally Sensitive Area (ESA) and NSA schemes (Chambers *et al.*, 2000). Solutions to the reduction of P in run-off have been shown to be effective in a range of countries and cropping and climatic conditions. Management techniques and potential benefits are shown in Table 3.7.

In summary, if reduction in the loss of P from agricultural land via run-off is to be effective on a small scale there is a need to:

- 'maximise' crop cover;
- have minimal cultivation;
- crop across not up and down slopes;
- reduce tillage effects that will cause soil dispersion;
- apply fertilisers to non-wet soils and when rainfall event is not imminent;
- reduce field drain efficiency so prolonging contact time, chance of sorption and reducing subsurface leaching;
- balance and manage P inputs and outputs for livestock and crops (Catt *et al.*, 1998; HGCA, 2000; MAFF, 2000b).

Table 3.7 Some practical management techniques for the reduction of P loss (modified and adapted from Pierzynski *et al.*, 2000).

Management technique	Benefits
Conservation tillage	Normally leaves at least 30% of the surface covered by plant residues, includes stubble mulch tillage, reduced tillage etc. reduces erosion and run-off
Contour cultivation	Cultivation follows contours – slows or stops downhill water movement
Vegetation soil cover	Intercepts rainfall (<45%) reducing impact energy and promoting infiltration – Important to achieve canopy closure in minimum time
Crop rotation	Planting a crop soon after previous one is harvested minimises exposure of bare soil
Companion crops	Crops which grow quickly planted in tandem with main crop – gives the benefits of vegetation soil cover
Vegetated waterways	Reduce velocity of run-off, prevent gully erosion, concentrates water in non-erodable channels
Cover crops	Crops that grown in periods when no crop is in soil (winter), not sold, but grow quickly – cover bare soil
Buffer strips	Used primarily for the control of water pollution, sited next to waterways and drainage ditches – vegetated zones (Box 3.3)

While an increase in the retention of P may occur by reducing agricultural runoff through management techniques that promote water infiltration this can have a considerable negative effect upon attempts to reduce NO_3^- leaching. Therefore, a balance needs to be struck on a site-specific basis as to the priority environmental concern and the potential impacts on the other macronutrient.

Box 3.3 Buffer zones and constructed wetlands

Buffer zones and constructed wetlands may not reduce soil erosion or leakage of P from agricultural land, but they will act as filters to protect surface waterways. They are often sited adjacent to waterways or zones that are highly sensitive to potential runoff and are basically uncultivated land between fields and waterways. The primary mechanism by which they are effective is by causing particulate material in runoff to drop out of solution by sedimentation. The vegetation in the buffer zone or wetland reduces overland flow velocity, increases the surface roughness and decreases the sediment carrying capacity. P sorption and plant uptake may also play a role in reducing P concentrations in runoff, although this is probably minor in the short-term (Uusi-Kämppä *et al.,* 2000).

The vegetation characteristics in the zone are crucial, as surface roughness needs to be at its maximum when the transport potential via run-off is also at its greatest. This does not often occur as high rainfall and run-off occur primarily in winter months when the vegetation cover may be minimal (Heathwaite, 1997).

The significance of livestock grazing on riparian land can be considerable, especially if watering or feeding areas are also located in the riparian zones. (Heathwaite, 1997). However, the *Code of Good Agricultural Practice for the Protection of Water* recommends the leaving of a 10-m buffer strip either side of watercourses and agricultural land (MAFF, 1998).

A range of options exists for construction of zones to reduce P loss from agricultural land, including wetlands, ponds and grassed and wooded strips. The effectiveness of these zones will depend to a certain extent upon the form of P fertiliser applied. Therefore, if slurries, which may have a large dissolved P load, are transported via run-off through a buffer zone, one may expect only a limited reduction in P load (Heathwaite, 1997).

Studies on small mixed farming catchments in Scandinavia have shown that grassed buffer zones may retain between 27 and 97% of total P load of which the majority was in particulate, whereas constructed wetlands and ponds reduced total P loads by only 41 and 17% respectively. The effectiveness of a range of buffer zones were observed to increase considerably with increasing width (Uusi-Kämppä *et al.,* 2000).

Nevertheless, catchment hydrology and the presence of preferential subsurface flow pathway, such as drains, will reduce the effectiveness of buffer strips, and therefore remedial efforts should still be directed at the agricultural source of P.

REFERENCES

Barber, S.A. (1995) *Soil Nutrient Bioavailability; A Mechanistic Approach*. 2nd Edition. Wiley, New York.

Brookes, P.C., Heckrath, G., De Smet, J., Hofman, G. and Vanderdeelen, J. (1997) Losses of phosphorus in drainage water. In: *Phosphorus Loss from Soil to Water*. (Eds H. Tunney, O.T. Carton, P.C. Brookes and A.E. Johnston). CAB International, Wallingford, pp. 253–271.

Brouwer, F.M., Godeschalk, F.E., Hellegers, P.J.G.J. and Kelholt, H.J. (1995) *Mineral Balances at Farm Level in the European Union*. Agricultural Economics Research Institute (LEI-DLO), The Hague.

Catt, J.A., Howse, K.R., Farina, R., Brockie, D., Todd, A., Chambers, B.J., Hodgkinson, R., Harris, G.L. and Quinton, J.N. (1998) Phosphorus losses from arable land in England. *Soil Use and Management* **14**, 168–174.

Chambers, B.J., Garwood, T.W.D. and Unwin, R.J. (2000) Controlling soil water erosion and phosphorus losses from arable land in England and Wales. *Journal of Environmental Quality* **29**, 145–150.

Cooke, G.W. (1981) The fate of fertilisers. In: *The Chemistry of Soil Processes*. (Eds D.J. Greenland and M.H.B. Hayes). John Wiley and Sons, Chichester, pp. 563–592.

Correll, D.L. (1998) The role of phosphorus in the eutrophication of receiving waters: a review. *Journal of Environmental Quality* **27**, 261–266.

Cunningham, J.E. and Kuiack, C. (1992) Production of citric and oxalic acids and solubilization of calcium phosphate by *Penicillium bilaii*. *Applied Environmental Microbiology* **58**, 1451–1458.

Daniel, T.C., Carton., O.T. and Magette, W.L. (1997) Nutrient-management planning. In: *Phosphorus Loss from Soil to Water* (Eds H. Tunney, O.T. Carton, P.C. Brookes and A.E. Johnston). CAB International, Wallingford, pp. 297–309.

Daniel, T.C., Sharpley, A.N. and Lemunyon, J.L. (1998) Agriculture phosphorus and eutrophication: A symposium overview. *Journal of Environmental Quality* **27**, 251–257.

Edwards, A.C. (1996) Soluble and particulate phosphorus losses for contrasting soil types and land use situations. *Soil Use and Management* **29**, 223.

Edwards, A.C. and Withers, P.J.A. (1998) Soil phosphorus management and water quality: a UK perspective. *Soil Use and Management* **14**, 124–130.

Edwards, A.C., Twist, H. and Codd, G.A. (2000) Assessing the impact of terrestrially derived phosphorus on flowing water systems. *Journal of Environmental Quality* **29**, 117–124.

Environment Agency (2000a) Inland Water Quality and Use. In: *Digest of Environmental Statistics*. http://www.environment.detr.gov.uk/des/index.htm

Environment Agency (2000b) Aquatic Eutrophication in England and Wales: A Management Strategy. Department of the Environment, Transport and the Regions.

Foy, R.H. and Bailey-Watts, A.E. (1998) Observations on the spatial and temporal variation in the phosphorus status of lakes in the British Isles. *Soil Use and Management* **14**, 131–138.

Gaynor, J.D. and Findlay, W.I. (1995) Soil and phosphorus loss from conservation and conventional tillage in corn production. *Journal of Environmental Quality* **24**, 734–741.

Gibson, C.E. (1997) The dynamics of phosphorus in freshwater and marine environemnts. In: *Phosphorus Loss from Soil to Water*. (Eds H. Tunney, O.T. Carton, P.C. Brookes and A.E. Johnston). CAB International, Wallingford, pp. 119–135.

Glendinning, J.S. (1999) *Australian Soil Fertility Manual*. CSIRO Publishing, Collingwood, Australia.

Harper, D.M. (1992) *Eutrophication of Freswaters: Principles, Problems and Restoration*. Chapman and Hall, London.

Haygarth, P.M., Chapman, P.J., Jarvis, S.C. and Smith, R.V. (1998) Phosphorus budget for two contrasting grassland farming systems in the UK. *Soil Use and Management* **14**, 160–167.

Heathwaite, A.L. (1997) Sources and pathways of phosphorus loss from agriculture. In: *Phosphorus Loss from Soil to Water* (Eds H. Tunney, O.T. Carton, P.C. Brookes and A.E. Johnston). CAB International, Wallingford, pp. 205–223.

Heckrath, G., Brooks, P.C., Poulton, P.R. and Goulding, K.W.T. (1995) Phosphorus leaching from soils containing different P concentrations in the Broadbalk experiment. *Journal of Environmental Quality* **24**, 904–910.

HGCA (2000) *P & K Fertiliser Planning; Using Soils and Plant Analysis to Plan Fertiliser Use*. Home-Grown Cereals Authority, London. Autumn 2000.

Hooda, P.S., Moynagh, M. and Svobada, I.F. (1996) A comparison of phosphate losses in drainage water from two different grassland systems. *Soil Use and Management* **12**, 224.

Hooda, P.S., Moynagh, M., Svoboda, I.F., Edwards, A.C., Anderson, H.A. and Sym, G. (1999) Phosphorus loss in drainflow from intensively managed grassland soils. *Journal of Environmental Quality* **28**, 1235–1242.

Hotsma, P. (1997) Present and future Dutch regulations to reduce phosphorus loss to water from agriculture. In: *Phosphorus Loss from Soil to Water*. (Eds H. Tunney, O.T. Carton, P.C. Brookes and A.E. Johnston). CAB International, Wallingford, pp. 410–412.

Johnes, P.J. and Hodgkinson, R.A. (1998) Phosphorus loss from agricultural catchments: pathways and implications from management. *Soil Use and Management* **14**, 175–185.

Laegreid, M., Bøckman, O.C. and Kaarstad, O. (1999) *Agriculture Fertilisers and the Environment*. CAB International, Wallingford, Oxfordshire, pp. 150–157.

Larsen, S. (1971) Residual phosphate in soils. In: *Residual Value of Applied Nutrients*. MAFF Technical Bulletin No. 20, pp. 34–40.

Lennox, S.D., Foy, R.H., Smith, R.V. and Jordan, C. (1997) Estimating the contribution from agriculture to the phosphorus load in surface water. In: *Phosphorus Loss from Soil to Water* (Eds H. Tunney, O.T. Carton, P.C. Brookes and A.E. Johnston). CAB International, Wallingford, pp. 55–75.

Mariën, F. (1997) European perspective on phosphorus and agriculture. In: *Phosphorus Loss from Soil to Water* (Eds H. Tunney, O.T. Carton, P.C. Brookes and A.E. Johnston). CAB International, Wallingford, pp. 329–337.

Mason, C.F. (1996) Water Pollution Biology. In: *Pollution, Causes, Effects and Control* (Ed. R.M. Harrison). 3rd Edition. The Royal Society of Chemistry, Cambridge, pp. 66–92.

MAFF (1997) *Controlling Soil Erosion – an Advisory Booklet for the Management of Agricultural Land*. MAFF Publications, MAFF, London.

MAFF (1988) *Fertiliser Recommendations*. Reference Book 209. HMSO, London.

MAFF (1998) *Budgets and Buffers*. R+D Newsletter, No. 4 Autumn 1998.

MAFF (2000a) *Predicting Phosphorus Lost from Agriculture to Water*. R+D Newsletter, No. 6 Summer 2000, p. 9.

MAFF (2000b) *Fertiliser Recommendations for Agricultural and Horticultural Crops*. MAFF Publications MAFF, London.

Morse, G.K., Lester, J.N. and Perry, R. (1993) *The Economic and Environmental Impact of Phosphorus Removal from Wastewater in the European Community*. Selper Publications, London.

Moss, B. (1998) *Ecology of Fresh Waters, Man and Medium, Past to Future*. 3rd Edition. Blackwell Science, Oxford.

Pierzynski, G.M., Sims, J.T. and Vance, G.F. (2000) *Soils and Environmental Quality*. 2nd Edition. CRC Press LLC, Boca Raton, Fla, pp. 155–207.

Reuter, D.J. and Robinson, J.B. (1997) *Plant Analysis, an Interpretation Manual*. 2nd Edition. CSIRO Publishing, Collingwood, Australia.

Rowell, D.L. (1994) *Soil Science, Methods and Applications*. Longman Scientific and Technical. Harlow.

Sakamoto, M. (1996) Eutrophication. In: *Water Resources, Environmental Planning, Management and Development* (Ed. A.K. Biswas). McGraw-Hill, New York, pp. 297–379.

Schlesinger, W.H. (1997) *Biogeochemistry, an Analysis of Global Change*. 2nd Edition. Academic Press, San Diego, pp. 383–401.

Sharpley, A.N. (1980) The enrichment of soil phosphorus in runoff sediments. *Journal of Environmental Quality* **9**, 521–526.

Sharpley, A.N., Daniel, T.C. and Edwards, D.R. (1993) Phosphorus movement in the landscape. *Journal of Production Agriculture* **6**, 492–500.

Sharpley, A., Foy, B. and Withers, P. (2000) Practical and innovative measures for the control of agricultural phosphorus losses to water: an overview. *Journal of Environmental Quality* **29**, 1–9.

Sharpley, A.N. and Halvorson, A.D. (1994) The management of soil phosphorus availability and its impact on surface water quality. In: *Soil Processes and Water Quality* (Eds R. Lal and B.A. Stewart). CRC Press, Inc, Boca Raton, Fla, pp. 7–90.

Sharpley, A.N., Smith., S.J. and Naney, J.W. (1987) Environmental impact of agricultural nitrogen and phosphorus use. *Journal of Agricultural Food Chemistry* **35**, 812–817.

Smith, K.A., Chalmers, A.G., Chambers, B.J. and Christie, P. (1998) Organic manure phosphorus accumulation, mobility and management. *Soil Use and Management* **14**, 154–159.

Steén, I. (1997) A European fertiliser industry view on phosphorus retention and loss from agricultural soils. In: *Phosphorus Loss from Soil to Water*. (Eds H. Tunney, O.T. Carton, P.C. Brookes and A.E. Johnston). CAB International, Wallingford, pp. 311–328.

Tisdale, S.L, Nelson, W.L., Beaton, J.D. and Halvin, J.L. (1993) *Soil Fertility and Fertilisers*. 5th Edition. Prentice Hall, New Jersey.

Tunney, H. (1992) Some environmental implications of phosphorus use in the European Community. In: *Phosphorus, Life and Environment, from Research to Application*. Proceedings of the 4th International IMPHOS Conference, Ghent, Belgium. World Phosphate Institute, Casablanca, Morocco, pp. 347–359.

Tunney, H., Breeuwsma, A., Withers, P.J.A. and Ehlert, P.A.I. (1997) Phosphorus fertiliser strategies: present and future. In: *Phosphorus Loss from Soil to Water* (Eds H. Tunney, O.T. Carton, P.C. Brookes and A.E. Johnston). CAB International, Wallingford, pp. 177–203.

Uusi-Kämppä, J., Braskerud, B., Jansson, H., Syversen, N. and Uusitalo, R. (2000) Buffer zones and constructed wetlands as filters for agricultural phosphorus. *Journal of Environmental Quality* **29**, 151–158.

Valk, H., Metcalf, J.A. and Withers, P.J.A. (2000) Prospects of minimizing phosphorus excretion in ruminants by dietary manipulation. *Journal of Environmental Quality* **29**, 28–36.

Van der Molen, D.T., Breeuwsma, A., Boers, P.C.M. and Roest, C.W.J. (1997) Dutch policy towards phosphorus losses in agriculture. In: *Phosphorus Loss from Soil to Water* (Eds H. Tunney, O.T. Carton, P.C. Brookes and A.E. Johnston). CAB International, Wallingford, pp. 407–409.

White, R.E. (1997) *Principles And Practice Of Soil Science, The Soil as a Natural Resource*. 3rd Edition. Blackwell Science Ltd, Oxford.

Whitelaw, M.A., Harden, T.J. and Bender, G.L. (1997) Plant growth promotion of wheat inoculated with *Penicillium radicum* sp. nov. *Australian Journal of Soil Research* **35**, 291–300.

Wild, A (1981) Mass flow and diffusion. In: *The Chemistry of Soil Processes* (Eds D.J. Greenland and M.H.B. Hayes). John Wiley and Sons, Chichester, pp. 37–80.

Withers, P.J.A. (1996) Phosphorus cycling in UK agriculture and implications for water quality. *Soil Use and Management* **12**, 221.

Withers, P.J.A., Davidson, I.A. and Foy, R.H. (2000) Prospects for controlling non-point phosphorus loss to water: a UK perspective. *Journal of Environmental Quality* **29**, 167–175.

Withers, P.J.A. and Jarvis, S.C. (1998) Mitigation options for diffuse phosphorus loss to water. *Soil Use and Management* **14**, 186–192.

Withers, P.J. and Sharpley, A.N. (1995) Phosphorus fertilisers. In: *Soil Amendments and Environmental Quality* (Ed. J.E. Rechcigl). CRC Pres, Inc, Boca Raton, Fla, pp. 65–107.

Soil Erosion 4

4.1 INTRODUCTION

The erosion of soil by water and wind is linked intimately with the cultivation of land, which can be traced back to at least 5000 BC (Goudie, 2000). Natural rates of erosion tend to be highest in semi-arid regions of the world, where sparse vegetation cover and intense climatic conditions combine to generate significant erosive potential. Historically, rates of soil loss by erosion have been lower in temperate and tropical regions, where more continuous vegetation cover reduces the erosive potential of heavy rainstorms or strong winds. The repeated cultivation of land for crop production and the removal of protective forest cover have resulted in soil becoming more vulnerable to erosion. The consequences of these losses can be severe not only for the productive capacity of the land, but also for habitats and environments downstream or downwind where products of erosion are deposited.

The impact of soil erosion within individual fields that have been affected by run-off or wind-blow is obvious, and has been the subject of much research activity since the 1940s, particularly in the United States (Morgan, 1995; Hudson, 1995). Off-farm impacts have received less attention, but are now recognised as an equally important component of the sediment loss issue. This chapter is subdivided into three sections that focus on the main causes of soil erosion from agricultural land, the resultant impacts on aquatic and terrestrial environments, and strategies that can be adopted to mitigate against soil erosion. The impacts associated with nutrients and pesticides lost from agricultural land in run-off are described in Chapters 2 (Nitrates), 3 (Phosphates) and 7 (Pesticides). In this chapter, the processes, extent and management of sediment loss by soil erosion are considered.

4.2 VULNERABILITY OF SOIL TO EROSION AND DEGRADATION

Extent of erosion

The serious widespread consequences of erosion from agricultural land were demonstrated in the South Western United States during the 1920s and 1930s, a period that is still referred to as the 'dust-bowl years' (Tivy, 1990). The repeated cultivation of semi-arid temperate grassland lead to accelerated erosion by wind and water. Soils degraded and lost their productive potential, while wind blown sediment was transported many hundreds of kilometres from the source. Despite the introduction of soil conservation strategies in many parts of the world particularly vulnerable to erosion, notably the semi-arid climatic zones, soil loss by erosion and consequent pollution continue to be a major environmental concern. In the United States, the Agricultural Research Service estimated total erosion losses of soil from cultivated land in 1992 of 3100 Mt year^{-1} (USDA, 2001). Current rates of erosion average 8.5 t ha^{-1}, which has lead to the US Environmental Protection Agency classifying siltation and nutrient enrichment as the major pollutant of freshwater in the United States, affecting 45% of rivers and lakes (NRCS, 1997). About 90% of US cropland is currently losing soil above the sustainable rate (i.e. that is faster than it is being formed). The Food and Agriculture Organisation (FAO) estimate that soil erosion rates in parts of Asia, Africa and South America are estimated to be about twice as high as in the USA. The FAO estimates that 140 million ha of high quality soil, mostly in Africa and Asia, will be degraded by 2010, unless better methods of land management are adopted (FAO, 2001).

In temperate regions sediment loss by erosion is perceived to be a less serious problem than in semi-arid environments. In the UK rates of soil loss are generally low, but have been increasing in recent years (Evans, 1996a). Successive reports by the Royal Commission on Environmental Pollution (RCEP, 1979, 1984, 1996) have drawn attention to the increasing environmental impact of sediment and nutrient losses by erosion. Soil surveys have shown that between 37 and 45% of arable land is at risk of erosion, and that on average 5% of that arable land will erode in any one year (Morgan, 1995). In a review of the impact of modern farming practices on the soil, Webb *et al.* (2001) noted that some soils, notably those on chalk parent materials or with sand/light loam texture, tended to be most vulnerable to erosion, but that in most cases crop yield was not affected by these losses, provided adequate fertiliser was applied. However, these authors did note that where erosion was occurring regularly, there was a need to reduce the off-farm impacts, such as silting of reservoirs and damage to fish spawning grounds.

Soils in upland areas are also vulnerable. Increased grazing pressure on the deep peat moors of Northern England, Wales and Scotland has lead to erosion and consequent loss of habitat and downstream impacts, such as water supply contamination (Evans, 1996a; Grieve and Hipkin, 1996). In some areas up to 20% of peat has been eroded, although it is difficult to distinguish human and natural causes of this soil loss. The RCEP 19th Report 'The Sustainable Use of Soil' concluded that erosion is a problem in soil upland areas of the UK, where it can have serious implications for conservation of habitats or plant communities. Erosion of arable land was not identified as a major national problem, nor was it seen to be seriously affecting the productivity of agricultural land. However, loss of soil particles carrying nutrients and pesticides were identified as a significant cause of environmental damage beyond the farm boundary (RCEP, 1996).

Elsewhere in Europe, soil erosion is also a serious problem, especially in southern Europe. It is caused by a combination of climatic conditions, steep slopes, thin vegetation cover and agricultural practices that leave the soil exposed to erosive agents. The areas with the greatest severity of soil loss due to both wind and water erosion are the Mediterranean, the Balkan Peninsula and the countries surrounding the Black Sea. In Ukraine, for example, 41% (17 million ha) of agricultural land was subjected to water and wind erosion in 1996 (EEA, 2001). In Sweden, rill and gully erosion by water affects on average 7% of arable fields each year, with typical rates of soil loss averaging 0.8 t ha^{-1} $year^{-1}$, and ranging up to 120 t $ha^{-1}year^{-1}$ (Alstrom and Akerman, 1992). The extent of vulnerable soils in Western Europe was described in the soil erosion map produced on behalf of the European Union (De Ploey, 1989).

Processes and products of erosion

Erosion and sediment transfer is a natural process that progressively transforms landscapes at a rate dependent primarily upon climate and geology. Unmanaged landscapes change imperceptibly in human terms. Removal of the natural vegetation cover leads to accelerated erosion. The ratio of natural to accelerated erosion varies widely according to climatic controls. Table 4.1 illustrates the typical erosion rates under natural vegetation and cultivated land. Removal of the protective vegetation cover results in losses being accelerated by several orders of magnitude. Typical soil loss rates in the UK are 0.1–0.5 t $ha^{-1}year^{-1}$, which equates to a layer of soil approximately 0.01–0.05 mm deep. This is equivalent to the rate at which soils form in temperate climates, and can be described as a sustainable loss to which adjacent aquatic and terrestrial habitats are naturally adjusted. However, the topsoil lost by erosion is the most fertile, as it contains the main soil organic matter and nutrient reserves. Accelerated erosion induced by cultivation can lead to soil losses up to 20 t $ha^{-1}year^{-1}$ or greater. These losses are often very localised, and can therefore have a major impact on adjacent habitats, such as streams and rivers.

Table 4.1 Typical erosion rates under natural vegetation and cultivated land in selected countries (Morgan, 1995).

	Natural	Cultivated land
	t $ha^{-1}year^{-1}$	
USA	0.03–3	5–170
China	0.1–2	150–200
Ethiopia	1–5	8–42
UK	0.1–0.5	0.1–20

Soil erosion is the process by which sediment is detached and transported, usually by water or wind, from one location to another. The detachment and transport of material from one location can be quantified in terms of an erosion rate, and can be related to causal processes, either natural or human-induced. The impact of these sediments and associated chemical pollutants depends on the depositional environment, which may be many kilometres from the source. However, to develop effective management strategies, a clear understanding of erosion processes is essential, as it is the nature of the process, acting on a particular material, that controls the potential environmental impact.

Soil loss and sediment transport depends upon:

- erosivity – the potential of the rainfall or wind to erode the soil surface;
- erodability – the susceptibility of the soil to erosion.

Erosivity is controlled by natural climatic processes, while the erodability can be directly influenced by human activities associated with the cultivation of land. The two main erosive agents that operate on agricultural soils are water and wind. In both cases soil particles must first be detached from adjacent particles, prior to being transported (Box 4.1).

Water erosion

Particle detachment by raindrop impact depends upon the kinetic energy of the rain, KE, which is given by:

$$KE = 0.5\ mv^2$$

where m is the raindrop mass and v is the terminal velocity. As raindrop diameter and hence mass increases with intensity, the characteristics of the rainfall are important in controlling the likelihood of detachment. In UK conditions, erosivity of rainfall reaches a maximum at 25–50 mm h^{-1}, rates that are experienced in heavy conventional storms. Field-based monitoring of water erosion incidents has shown that regularly cultivated soils in the UK are being eroded by low intensity rainfall, in the range 5–10 mm h^{-1} (Evans, 1996a). Amounts of soil lost from individual fields can exceed 20 t ha^{-1} in severe localised cases. The impact of such events depends on the characteristics of the depositional environment (Figure 4.1). If immediate, localised deposition of fine

Figure 4.1 Rill erosion in a winter cereal crop.

sand and silt-sized sediment occurs in a watercourse, then the impacts on aquatic organisms can be dramatic. More usually, sediment disperses through the downstream channel network and as a result the environmental impact is more diffuse.

Particle transport by flow over the soil surface occurs when surface depression storage and the infiltration capacity is exceeded (Morgan, 1995). Flow is rarely uniform, but concentrates in shallow natural topographic depressions, or man-made channels, such as tractor wheelings. For particles larger than 0.2-mm diameter (fine sand sized and greater, the potential to transport detached particles is controlled by the velocity of the run-off and the particle diameter. Erosion risk is greater with higher water velocity and smaller particle size. Coarser sand and gravel-sized material moves less frequently, and tends to be redistributed within fields during an erosion event, as is evidenced by the fans of material deposited at the base of rills and gullies. For finer material, below 0.2-mm diameter, erosion potential depends on whether the particles derive from a cohesive (clay-rich) or non-cohesive soil, with the latter being more vulnerable to transport.

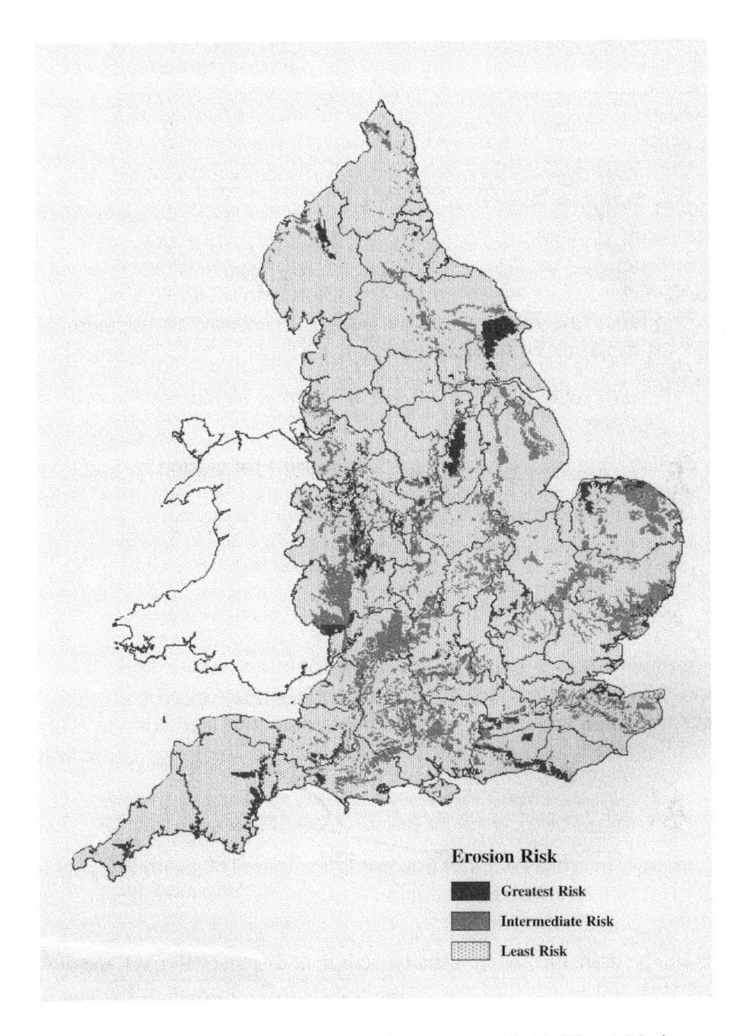

Figure 4.2 Soils at risk of water erosion in England (MAFF, 1999a).

The spatial distribution of soils vulnerable to water erosion reflects the importance rainfall characteristics, surface topography and soil texture. Figure 4.2 shows the soils at risk from water erosion in England. The highest risk soils are those containing a high proportion of sand and silt that occur on slopes >7°. The assessment of erosion risk, and methods for reducing the vulnerability of soils to erosion, such as maintaining a high soil infiltration rate and reducing the catchment area to minimise the risk of sediment associated run-off are discussed later in this chapter.

Box 4.1 Processes of erosion (Morgan, 1995)

The two major transport agents are water and wind. Erosion is a two-phase process: *detachment* of soil particles, followed by *transport*. When sufficient energy is no longer available to transport the particles, then *deposition* occurs.

Water

Detachment

Rainsplash, controlled by intensity of rainfall, soil characteristics (e.g. texture and organic matter content) and soil antecedent conditions. Coarse silt and fine sand-sized particles (0.05–0.20-mm diameter) are most easily detached by rain.

Overland flow, occurs when surface storage and infiltration capacity of the soil is exceeded. Flow velocity and soil coherence controls particle detachment rate. For example, clay particles are more resilient to detachment because of the cohesiveness of individual clay particles.

Transport

Once a particle has become entrained, it will be transported by water until the energy in the flow reduces below a threshold that is grain size dependent. Coarser, sand-sized particles are deposited first (closest to the source of the sediment) while silt and clay-sized material (<0.06-mm diameter) tends to remain in suspension for longer.

Wind

Detachment

The capacity of wind to detach particles depends upon the critical shear velocity of the wind, which is controlled by wind speed and surface roughness. Particles between 0.10 and 0.20-mm diameter are most susceptible to wind erosion.

Transport

Once detached in motion, sediment transport takes place in suspension (for particles <0.2-mm diameter) or saltation (bouncing). This saltation process will detach other particles. Deposition occurs when wind speed is reduced, for example by a barrier such as a hedge.

Wind erosion

Soil erosion by wind is most common in arid areas, and on coastlines, where winds are strong and vegetation is sparse due to salt deposition. The severity of wind erosion depends on wind speed and soil/vegetation surface roughness (Box 4.1). Once soil particles have become detached and are in motion, they can be transported considerable distances. A sequence of drought years in the 1930s in the South West USA resulted in the transport of fine sand and silt-sized particles more than 1000 km from the source of the erosion (Goudie, 2000). The resulting sedimentation is regarded as one of the most widespread forms of water pollution (Tivy, 1990). As with water erosion, the disaggregation and transport of surface soil horizons by wind removes particulate material that includes organic, nutrient-rich components associated with the finer fractions of the eroded material. These fractions tend to fall within the range 0.05–0.5-mm diameter (Morgan, 1995).

In the UK, the more temperate climate and surface topography tends to restrict the area of soils vulnerable to wind erosion. Mostly, it is the fine sandy and peat soils in arable cultivation that are vulnerable to loss. Most 'wind-blows' tend to occur in the spring and early summer, and affect fields planted to high value crops such as sugar beet, onions and carrots (Evans, 1996a; MAFF, 1998). Estimates of the spatial extent and quantity of soil lost during periods of wind erosion are necessarily tentative, due to the difficulties in measuring deposition rates across large areas downwind of a 'wind-blow'. Evans and Cook (1986) observed moderate and severe blows in the Cambridge peat fens and Nottinghamshire sandlands in 5 or 6 years in 10, with soil loss rates of 5–10 $m^3 ha^{-1}$ being typical.

Agricultural land management practices and accelerated erosion

Many land management practices have been found to increase erosion risk. Often these are integral with the progressive intensification of agriculture that has occurred in recent decades. Globally, it is acknowledged that the progressive increase in area of cultivated land during the twentieth century has exposed more land, often unsuitable for arable cultivation, to the erosive forces of water and wind (Parkinson, 1995; Brown *et al.*, 2000).

Agricultural land management practices that are known to increase the vulnerability of soils to erosion are given in Table 4.2. In many cases, it is the combination of a number of these practices with naturally erodible soils or steep slopes that leads to enhanced risk of serious erosion. Repeated cultivation changes the soil configuration, and has been shown to reduce the organic matter content of soils such that structure stability can be compromised. Surveys conducted by the National Soil Resources Institute in England and Wales have revealed that soil organic matter levels in arable and ley-arable cropping have declined by an average 0.5% over the 15-year period 1980–1995 (MAFF, 2000). For sandy soils that tend to have a naturally low organic matter content, the continued loss of organic matter can lead to capping of the soil surface on rainfall impact. This reduces infiltration and allows runoff to occur more frequently. For a detailed review of soil factors influencing accelerated water erosion, see Evans (1996b).

Table 4.2 Land management practices which increase vulnerability of soils to erosion.

Agricultural system and cultural practice	Impact
Primary and secondary cultivation in continuous arable rotations	Structural deterioration, decline in organic matter content, reduction in surface roughness, leading to increased vulnerability to autumn/winter rainfall
Cultivating up and down slope	Increased opportunity for rill and gully establishment

Crop residue removal or burning	Decline in organic matter content
Cultivation of crops with low ground cover in critical periods	Increased duration of risk period for crops such as winter wheat and oilseed rape (autumn), potatoes and sugar beet (late spring), horticultural crops (e.g. onions)
Fertilisers as primary nutrient source, not manures	Declining organic matter content (MAFF, 1998), reduced structural stability
Increased machinery size and use of tramlines	Compaction of soils and reduced infiltration rate (if low ground pressure tyres not fitted)
Drainage of peat soils	Desiccation and oxidation of organic-rich soils, increased risk of 'wind-blows'
Hedge removal and field enlargement	Increased catchment area for developing rills and gullies
Grassland, increased stocking rates	Destruction of protective vegetation layer, increased trampling damage by animals, reduced infiltration rates

Crop rotation exerts a significant influence on susceptibility to erosion. Evans (1996a) notes that the shift from spring to winter cereals in the UK since the 1970s, encouraged by changing EU price support mechanisms, has lead to an increase in incidence of within-field river bank and ditch erosion, accompanied by the siltation of more gently sloping reaches of rivers. Cultivation and drilling of autumn sown crops such as winter wheat and oilseed rape leaves the soils with a fine surface during the autumn period. Late drilling and heavy autumn rainfall have lead to the increased incidence of water erosion in recent years in the UK (Davidson and Harrison, 1995). Similarly, spring sown crops such as potatoes and sugar beet have been implicated in serious erosion of vulnerable soils. In a survey of erosion and farming practices in England and Wales, Skinner and Chambers (1996) asked farmers to consider which land management practices would affect the severity of soil erosion. The factors identified in order of priority were arable cropping (20%), tractor wheelings (16%) and tramlines (8%). Removal of hedges, which leads to an increased catchment area for developing rills and gullies has also been found to be important (MAFF, 1998, 1999b). In the United States, continuous maize production in the absence of soil conservation measures has made a significant contribution to soil loss from agricultural land (NRCS, 1997).

Ultimately, it may be that major changes in methods of cultivation and the intensity of land use are needed to reduce rates of erosion from agricultural land. Reganold *et al.* (1987) reported on the long-term effects of conventional and organic farming on soil productivity, depth and erosion rates on soils in Washington State, USA. Silt loam soils on two adjacent farms were cultivated with winter cereals for 37 years. At the end of the period the topsoil of the organically farmed soil was 16 cm deeper than the conventionally farmed soil. The differences were attributed to significantly greater losses of soil by erosion on the conventionally farmed soil. Measured losses on the conventional field in this area, where soil erosion rates were high, averaged 32 t ha^{-1}, while those from the organically farmed soil were four times lower, typically 8 t ha^{-1}. It was predicted that at the current rate of erosion, all the topsoil on conventional farms growing cereals would be lost by the third decade of the twenty-first century.

4.3 ENVIRONMENTAL IMPACTS OF SOIL EROSION

Impacts of sediment on aquatic environments

The transfer of soil particles eroded from farmland to adjacent watercourses can have significant impacts on water quality. Some impacts, primarily those of a physical nature, are transient, as erosion is an episodic process, linked to infrequent, high magnitude climatic events. Other impacts, associated with chemical pollutants, tend to have a cumulative in effect on aquatic organisms. Most authorities (for example, NRA, 1992) classify these impacts as diffuse pollution, as major erosion events tend to affect large areas of land within a given catchment. Successful strategies to control losses and mitigate against these impacts tend to be catchment based (Napier, 1990).

Physical impacts

Erosion generates particulate material, predominantly of an inorganic nature, that can have deleterious impacts on both plant and animal life. Despite the fact that erosion events are short-lived, the impacts can be very persistent. Clearly, the frequency of storm events controls the ability of an aquatic ecosystem to re-adjust to less sediment rich conditions after run-off events. Physical impacts on aquatic organisms tend to be difficult to assess, as periods of high suspended sediment concentration are accompanied by oxygen depletion and elevated concentrations of chemical pollutants, such as pesticides bound to silt and clay. Example impacts of lowland rivers in the UK are given in Table 4.3.

Table 4.3 Physical impacts of increased sediment load on aquatic organisms (Leeks, 1995; Evans, 1996a).

Organism	Impact
Invertebrates (mayflies, stoneflies)	Thin films of fine sediment can inhibit or eliminate populations. Some species better adapted to poor water quality (e.g. midge larvae) may increase in numbers.
Fish	Silt can smother spawning beds, by infilling the matrix of coarse gravel stream bed. This reduces oxygen availability and kills fish eggs. This restricts number of suitable breeding areas. Salmonids particularly effected Reduction of number of invertebrates as food supply. Sub-lethal gill damage leading respiration problems.
Algae and macrophytes	Progressive reduction in number and diversity with increasing suspended sediment concentration. Increase in population of coarser weeds such as *Cladophora* at the expense of finer-leaved macrophytes.

The impacts of soil erosion can be very localised. For example, at a confluence between a small ditch or stream which is supplying the sediment, and a larger river inhabited by a wide range of aquatic organisms. Sand-sized sediment (<0.06-mm diameter) tends to deposit close to the source of the erosion, where the energy of the flow medium declines below the critical threshold necessary to maintain the sand in suspension. Finer silt and clay-sized sediment can be transported many kilometres from the sediment source and present a pollution risk to rivers and estuaries downstream. Ironically, the impact of sedimentation in gravel-bedded rivers can be exacerbated by drought after a period of flooding (Leeks, 1995). Dry periods and low flow can cause the consolidation of the bed matrix, resulting in persistent long-term difficulties for spawning fish.

Long-term studies such as that conducted by Heaney *et al.* (2001) demonstrate that declines in the freshwater survival of salmonids may at least in part be attributed to diffuse inputs of fine sediment which smothers salmon spawning redds. Susceptibility of juvenile salmon during the egg to smolt stages is particularly evident; the deposition of sediment creates anaerobic conditions that favour nitrate and ammonia production, which is particularly toxic to salmonids (Massa *et al.*, 2000). Sediment also reduces oxygen availability to eggs buried within the redds which also contributes to mortality (Soulsby *et al.*, 2001). Hence, changes to the management of agricultural land is necessary if this problem is to be reduced. As well as attempts to improve the management of soil to minimise sediment losses, organisations such as the Environment Agency also conduct remedial actions (for example, by 'ploughing' spawning areas inundated with sediment) to improve spawning success.

Reduction in water quality and the consequent decline in fish numbers due to persistent accelerated erosion have impacts on economically important fisheries (Evans, 1996a). Water purification and treatment costs increase as sediment concentration increases, but it is not possible to identify the proportion of the costs that can be attributed to the removal of sediment as opposed to other treatment costs. Evans (1996a) noted that the water industry in England and Wales invested £1.8 bn between 1991 and 1996 in drinking water treatment, and postulated that if only 1% of these costs were attributed to treatment to remove excess sediment in the water, that amounted to £3.6 m year^{-1}. Young and Osborn (1990) noted that the Conservation Reserve Program (CRP) in the United States in the 1970s and 1980s produced economic benefits estimated between US$6 bn and US$13 bn. The principal benefit was the reduction in off-farm impacts of erosion, which were cut by 25%. In particular, surface water quality increased significantly, reducing water treatment costs and eutrophication/sediment problems. There will be additional, largely unquantified water supply costs in catchments where reservoirs are used to store water. Sedimentation at rates greater than that predicted at the reservoir design stage due to greater rates of soil erosion will decrease the effective life-span of the reservoir and increase the water purification costs.

Chemical impacts

The main chemical pollutant impact of sediment derived from erosion of agricultural land relates to those materials bound to soil particles that desorb in receiving freshwaters. These chemical species include macronutrients, notably phosphorus, that contribute to eutrophication, and pesticides. Previous studies have demonstrated the close link between phosphorus losses from agricultural land and the extent of surface runoff and erosion (Heathwaite, 1997). Phosphorus is strongly bound to clays and sequioxides, and is an integral component of soil organic matter (Chapter 4). Disaggregation of soil particles leads to preferential transport of clay, silt and finely divided organic matter, that contains bound particulate P. Studies in small catchments in Denmark showed that between 45 and 82% of the P transported to watercourses can be in particulate form, equating to 0.05–0.47 kg P ha^{-1} (Kronvang, 1990; Hasholt, 1991). Sharpley and Smith (1990) reported that particulate P accounted for an average 85% of P transported from cropped watersheds in South West USA. Run-off derived particulate P typically accounted for 90% of the total P losses from arable land observed by Catt *et al.* (1994) in experiments conducted in southern England.

Once sediment-bound P has been transported and deposited in river channels, the ecological impact will be dependant upon the rates of P desorption. This process is complex, both physically and chemically, and is discussed in detail by Gibson (1997). Physical disturbance of sediments, for example during subsequent flood episodes, or by human activity, will stimulate the release of soluble P that is bio-available to algae and diatoms. Changes in the chemical environment in rivers and lakes, notably the redox potential and the availability of free Fe and Ca to form complexes, will have a marked impact on P release rates. Actual release rates are very variable, but sediments containing in excess of 1 mg Pg^{-1} dry sediment are liable to release P that has been transported into the waterbody bound to the sediment (Gibson, 1997).

The transfer of pesticides bound to sediment eroded from agricultural land is another serious impact of accelerated soil loss. Routine analysis of drinking water is conducted throughout Europe in order to assess compliance with standards set by the European Commission. Maximum admissible concentrations for potable water should not exceed 0.1 $\mu g\, l^{-1}$ for a single pesticide and 0.5 $\mu g\, l^{-1}$ for all pesticides (Chapter 7). Residual herbicides such as the triazine group are designed to bind to soil particles and then kill germinating seedlings during the early phases of crop establishment. Experimental evidence describing the magnitude of pesticide loss from agricultural land directly associated with surface runoff and soil erosion is restricted, as few sampling programmes are adapted to sample sediment being transported from fields during runoff events. In a study conducted on arable land in Herefordshire, UK, Matthiessen *et al.* (1992) observed concentrations for individual herbicides up to 680 times higher than under background, low flow conditions.

The increased frequency of erosion from forage maize crops, particularly in south west England has lead to concerns expressed by the Environment Agency. The area of forage maize grown in south west England has increased six-fold between 1988 and 1995 (Environment Agency, 2001). Surveys of water quality indicate that although the total amount of the herbicide atrazine, which is used widely on maize seedbeds, has declined in recent years, it still forms a significant proportion of the total pesticide load. The Environment Agency also note that land used for growing maize is also at risk from erosion because fields are often left compacted and bare over winter.

Horticultural crops can be particularly prone to loss of pesticides bound to sediment due to low ground cover and high rates of pesticide applications. Harrod (1994) measured concentrations of the organochlorine insecticides aldrin and deldrin in excess of 1000 $\mu g\, kg^{-1}$ in sediment rich run-off from daffodil bulb fields in South West Cornwall, UK. Subsequent analysis of dieldrin levels in eels downstream the Newlyn River, Cornwall showed concentrations greater than 20,000 $\mu g\ l^{-1}$ fresh weight. These organochlorine pesticides have residence times in excess of 12 months, which accounts for the continued measurement of these persistent pesticides in sediments and 3 years after their use was banned in intensive horticultural production systems (Chapter 7).

Impacts of sediments on land

Although not regarded as pollution in the traditional sense, sediment losses from within fields and redeposition elsewhere within the terrestrial environment can have a significant local impact on environmental quality as well as having consequences for human health and the local rural economy. Some of these physical impacts, which are very difficult to quantify, are summarised in Table 4.4. Severe erosion can lead to reductions in crop yields (for example, Catt *et al.,* 1994; Chambers and Davies, 1995) and increased reliance in inputs of fertilisers and other agrochemicals to maintain yields. Crops downslope of an eroded field may be smothered by sediment, hence retarding growth. The deposition of nutrient-rich sediment may impact on the ecology of surrounding non-farmed habitats. A number of off-farm impacts are listed in Table 4.4. These are not discussed further here, but further information may be obtained from Evans (1996a) and Boardman *et al.* (1990).

Table 4.4 Physical impacts of sediment transport and deposition on the terrestrial environment.

Location	Impact
On-farm, within field	Severe gullying – restriction of cultivations, disruption of drainage systems Crop smothering – crop retardation or loss
On-farm, adjacent to the field	Drainage ditch disruption – destabilisation (erosion) or blockage (sediment deposition)
Off-farm	Sediment deposition on roads, highways, and adjacent properties, blocked drains and gutters Nutrient enrichment of adjacent semi-natural habitats by sediment transported by wind or water Dust deposition, possible links to incidence of respiratory problems in humans and animals

4.4 PRACTICAL SOLUTIONS TO SOIL EROSION

The consequences of accelerated soil erosion can be severe, resulting in loss of the productivity capacity of soil on-farm, and major environmental degradation off-farm. Strategies to control soil loss and hence minimise pollution attributable to this source vary from very simple, cost-effective and locally applied measures to catchment-wide government support initiatives, supported by financial and technical aid. Morgan (1995) defined the aim of soil conservation 'to obtain the maximum sustained level of production from a given area of land whilst maintaining soil loss below a threshold level which, theoretically, permits the natural rate of soil formation to keep pace with the rate of soil erosion'.

In essence, the function of erosion control strategies is to reduce rates of sediment loss close to that which would occur under natural conditions. In theory, this will reduce loss of nutrients and pesticides to a minimum, decrease rates of sedimentation of watercourses and minimise terrestrial impacts. It is unlikely that accelerated erosion impacts can ever be completely halted in agricultural systems that utilise field scale cultivation methods to establish crops. The target is to reduce loss rates to the rate of soil formation, which in temperate climates average 0.1 mm year^{-1} (approximately equivalent to 1 t ha^{-1}, assuming a soil bulk density of 1 tm^{-3}). Sound strategies for control of erosion are based on detailed assessment of erosion risk, prior to the implementation of control measures at either the field or the catchment scale.

Erosion control strategies can operate at different scales and utilise contrasting approaches. Agronomic and soil management measures control soil detachment and transport, operate at the field and farm scale, and through a concerted regional or catchment co-ordinated approach can be applied across many farms in vulnerable areas. Several of these approaches are described here. Mechanical or physical methods will control transport of sediment rather but do little to prevent detachment, and tend to be expensive solutions to erosion problems. Examples of mechanical methods include bunding and terracing, which are described in detail by Hudson (1995) and Morgan (1995). This section focusses on erosion risk assessment and control using appropriate soil management and crop husbandry.

Assessment of erosion risk

The assessment of potential vulnerability to erosion is based on conducting a rigorous, farm scale evaluation (FSE) of site characteristics (MAFF, 1999a). In the UK, most advice is directed towards the control of water erosion. Key criteria are soil texture, slope and annual rainfall. A risk assessment should be based on a field by field examination, noting topographical features that might be of importance in controlling run-off, such as hollows where gullies might form. Table 4.5 gives the erosion risk for soils in areas of England and Wales where the rainfall exceeds 800 mm year^{-1}. Additional factors such as organic matter content, soil structure, land use history and local knowledge are important in assessing erosion potential. Once the risk has been assessed for individual fields, the preparation of a farm erosion risk assessment map will indicate where contiguous fields can increase the risk. For example, two high risk fields separated by a fence line will pose a greater risk of erosion due to the impact that the combined surface run-off could have on the topographically lower parts of the farm. Risk assessment definitions are given in Table 4.6. For land classed as high and very high risk to erosion, it is recommended that precautions be taken prior to the cultivation of those crops which are known to increase erosion potential.

Table 4.5 Erosion risk classes for areas of England and Wales where average annual rainfall exceeds 800 mm year^{-1} (MAFF, 1999a).

Soil texture	Steep slopes >7°	Moderate slopes 3–7°	Gentle slopes 2–3°	Level ground <2°
Sand, Loamy sand Sandy loam Sandy silt loam	Very high	High	Moderate	Slight
Silt loam Silty clay loam	High	Moderate	Lower	Slight
Other mineral soils	Lower	Slight	Slight	Slight

Field and crop based erosion control strategies

Evaluation of site susceptibility to erosion allows the adoption specific soil and crop management strategies that usually concentrate on high and very high risk land categories. The recommended cropping for these categories of land in England and Wales are shown in Table 4.7. If crops known to be associated with high erosion risk are to be grown, then appropriate soil and crop husbandry measures should be carried out prior to and during the growth of the crop.

Table 4.6 Field/soil erosion risk definitions (MAFF, 1999a).

Risk category	Definition
Very high	Rills are likely to form in most years and gulllies may develop in very wet periods
High	Rills are likely to develop in most seasons during wet periods
Moderate	Sediment may be seen running to roads, ditches or watercourses, and rills may develop in some seasons during very wet periods
Lower	Discoloured run-off may enter ditches or watercourses
Slight	Any water running of the site is unlikely to be discoloured

Table 4.7 Recommended cropping strategy for high and very high risk sites (after MAFF, 1999a).

On very high risk and high risk sites	Cropping
Avoid the following crops unless precautions have been taken prior to crop establishment	Late sown winter cereals, potatoes, sugar beet, field vegetables, outdoor pigs, grass reseeds, forage maize
The following can be carried out with care	Early sown winter cereals, oilseed rape, spring sown cereals, spring sown linseed, short rotation coppice
Establishment of these crops will reduce the overall erosion risk	Long grass leys, permanent grass, woodland (excluding short-term coppice)

There are a variety of management strategies that can be adopted once the vulnerability of a soil has been recognised and the appropriate choice of crop made for that site. Strategies tend to be adapted to particular climatic zones and agricultural systems. For UK cropping situations detailed guidelines are given by MAFF (1998, 1999a,b). These guidelines identify a number of good management practices for the control of water erosion:

- plan the type and timing of cultivations to minimise the period when the soil is left vulnerable to erosion. Fine seedbeds and bare land after root crop harvesting are noted as potential risks. Compaction at plough depth or in tramlines should be avoided in high risk situations;

- cultivating across the slope is noted as likely to reduce the risk of run-off on gentle slopes, but will have less effect on steeper, undulating topography;

- avoid the preparation of fine seedbeds for vulnerable soils, and if possible use non-inversion tillage methods that retain some residues from the previous crop on the soil surface;

- avoid deep ploughing to retain organic matter near the soil surface;

- increase the stability of topsoils by the regular use of bulky organic manures, taking care to avoid excessive amounts of nitrogen due to the nitrate leaching risk. Application of slurries can also help to increase organic matter content of soils, but the impact is less than for manures due to the lower DM content of slurry;

- consider adjusting the crop rotation to maximise the winter cover and extend the use of long grass leys;

- select varieties that can be harvested in a more timely manner to avoid wetter periods of the year and to allow more timely establishment of the next crop.

Specific recommendations for winter cereals are given as an example below.

- remove any compaction present before establishing the crop, and time operations to minimise risks of causing further compaction;

- where possible establish without ploughing and allow some chopped straw to be left on the surface;

- avoid overworking the soil; leave seedbeds as coarse as practicable and sow early enough to achieve a minimum of 25% crop cover before early winter;

- ideally, drill without tramlines, or avoid using tramlines until the spring and avoid rolling in the autumn, especially if the soil is wet;

- following harvest, leave land in stubble, preferable with a cover of chopped straw, until the next crop can be established.

Strategies to control wind erosion in the UK, based on MAFF (1999b), include:

- provision of shelter belts of trees or hedges on vulnerable sites to reduce the erosive potential of the wind. Shelter belts should allow 30–50% of the wind to pass through. The benefit of a shelter belt usually extends about 20 times its height downwind;

- provision of extra protection for establishing crops by use of nurse crops, such as winter barley or rye, or plant straw to increase soil surface roughness;

- use of mulches, for example paper sludge or sugar beet waste to help stabilise the soil surface;

- where possible increase the soil surface roughness, by using fewer cultivations.

Within field soil conservation and crop management strategies in tropical and sub-tropical regions is based on similar good management to that recommended for the UK, but with particular emphasis on maximising and maintaining ground cover (Hudson, 1995; Morgan, 1995). Crops that are grown in rows, tall tree crops and low-growing crops with large leaves are particularly vulnerable to erosion. For example, the continuous cultivation of maize and soybeans in the Cornbelt of the United States has produced serious, persistent erosion problems that have been effectively tackled through Soil Conservation Service programmes focussing on stubble-mulch cultivation (Napier, 1990). For successful implementation of erosion control measures, changes in crop husbandry practices must be technically sound, and socially and economically acceptable.

Farm and catchment based erosion control strategies

In order to be fully effective, erosion control strategies must operate at field, farm and catchment scale. Vulnerability to water and to a lesser extent wind erosion is topography-dependent. Farm scale approaches are based on understanding water and wind erosion concentration zones within the farmed landscape. MAFF (1999a) recommend the use of grassed buffer strips and set-aside land to reduce erosion risks in concentration zones such as valley floors, particularly at the base of steep slopes. These zones are also able to act as semi-natural filters to retain sediment that has mobilised upslope. In this way, the impact on off-farm environments can be minimised. Buffer strips are discussed in more detail in Chapter 3 in relation to the control of P losses from farmland. To effectively detain sediment from runoff, it is recommended that these strips be at least 20-m wide (Davies and Christal, 1996). Verstraeten and Poesen (1999) report on the use of earth-bunded retention ponds, usually <1 ha in area, to detain sediment generated by surface run-off in Belgium. These ponds need to be regularly cleaned out and the sediment returned to farmland in order to maintain their effectiveness.

Box 4.2 The US Conservation Reserve Programme (Napier, 1990; Young and Osborn, 1990; USDA, 2001).

The Conservation Reserve Programme was introduced in the early 1980s, and was designed to remove highly erodible land from agricultural production. Agents of the US Federal Government are authorised to enter into contractual agreements with landowners for removing land from production for a 10-year period. Compensation payments were made to farmers who could not use registered land for the production of food or fibre crops. Registered land had to have a continuous groundcover, which could be established with grant aid, but thereafter should be maintained by the landowner.

Landowners were allowed to register land that had been cultivated at least once during the period 1981–1985. Approximately 9 million hectares have been enrolled in the scheme, at a cost of $1.1 million year^{-1} for the late 1980s and early 1990s. Overall the programme has been estimated to have produced a net economic benefit of between $3400 and 11,000 m, 25% of which have been attributed to the reduction in water treatment costs.

Assessments of the environmental impact of the CRP have demonstrated that significant reductions in soil loss have been achieved. In Ohio, 60,000 ha have been enrolled with reductions in sediment losses averaging 37 t ha^{-1}. Much of this land is now eroding at rates of <2 t ha^{-1}. Average erosion rate on CRP-enrolled lands across the US has declined from 50 to 3 t ha^{-1}year^{-1}, and there has been an estimated reduction in erosion of 694 t year^{-1}.

The implementation of catchment and region-wide schemes, supported by government investment, has been shown to be very effective in reducing sediment loss rates. The US CRP is summarised in Box 4.2. This is an example of an approach to soil erosion that considers sediment pollution and the reduction of the productive potential of farmland to be a serious problem. There have been many examples of effective catchment-based erosion control schemes. Most of these rely on improving traditional systems instead of imposing new techniques from outside. Morgan cites an example from Mali, West Africa where serious erosion to land producing a range of arable crops and fodder for livestock. After a series of experimental measures were evaluated, it became clear that a combination of tree planting, contoured grass strips, stone bunds, diversion drains and some terracing provided more protection for the land. These measures were zoned across the landscape (see Figure 4.3). In this way, it is possible to integrate measures that when combined to minimise within-field erosion losses and detain any material that does become mobilised as close to the source as possible. Hence off-farm pollution risks are reduced to a minimum.

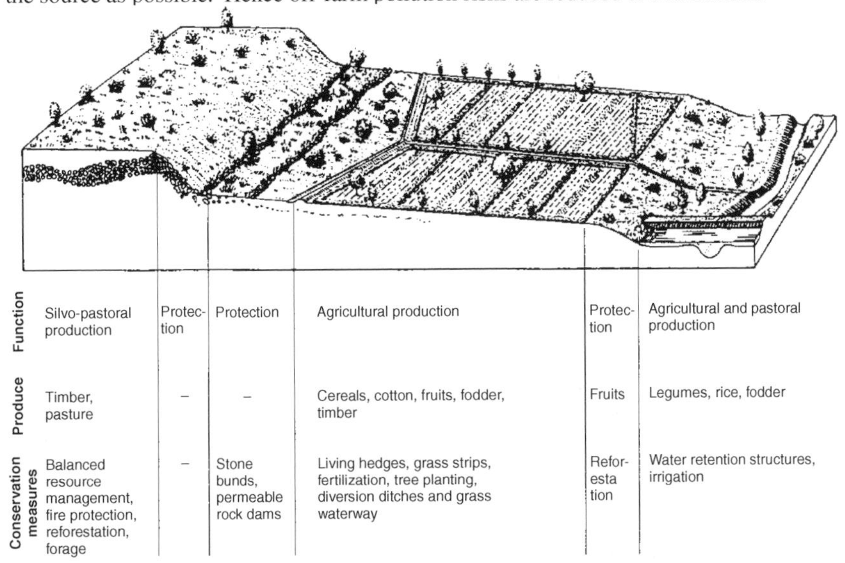

Figure 4.3 Catchment based soil conservation scheme on cultivated land in southern Mali (Hijkoop, van der Poel and Kaya, 1991, cited in Morgan, 1995).

The re-establishment of a permanent ground cover may be the only solution to persistent soil erosion problems, particularly in regions where the sediments impact upon sensitive habitats. Predictions are that in a changed climatic scenario of warmer, wetter winters for much of Western Europe, the incidence of soil erosion will increase (Boardman and Favis-Mortlock, 1993). The impact of climate change will be complex, but two important factors will directly affect erosion rates. Greater winter rainfall will increase the erosivity, while warmer temperatures will increase rates of organic matter loss and reduce soil structural stability, thus increasing erodability. Hence the need for effective erosion control strategies to control pollution associated with sediment from agricultural land will become more important in the future.

REFERENCES

Alstrom, K. and Akerman, A.B. (1992) Contemporary soil erosion rates on arable land in Southern Sweden. *Geografiska Annaler* **74A**, 101–108.

Boardman, J., Foster, I.D.L. and Dearing, J.A. (1990) *Soil Erosion on Agricultural Land*. Wiley, Chichester.

Boardman, J. and Favis-Mortlock, D.T. (1993) Climate change and soil erosion in Britain. *Geographical Journal* **159**, 179–183.

Brown, L., Flavin, C. and French, H. (2000) *State of the World 2000*. Worldwatch Institute, 262pp.

Catt, J.A., Quinton, J.N., Rickson, R.J. and Styles, P. (1994) Nutrient losses and crop yields in the Woburn erosion reference experiment. In: *Conserving Soil Resources: European Perspectives* (Ed. R.J. Rickson). CAB International, Wallingford, pp. 94–104.

Chambers, B. and Davies, D.B. (1995) The impact of water erosion on crop productivity and soil fertility in the UK. In: *Soil Management in Sustainable Agriculture* (Eds H.F. Cook and H.C. Lee), Wye College Press.

Davidson, D.J. and Harrison, D.J. (1995) The nature, causes and implications of water erosion on arable land in Scotland. *Soil Use and Management* **11**, 63–68.

Davies, D.H.K. and Christal, A. (1996) Buffer strips – a review. In: *Diffuse Pollution and Agriculture* (Eds T. Petchey, B.D'Arcy and A. Frost), SAC, Aberdeen.

De Ploey, J. (1989) *Soil erosion map of Western Europe*. Catena, Germany.

Environment Agency (2001) *Environment South West. A report on the state of the environment in the South West of England.*

EEA (2001) *Down to earth: Soil Degradation and Sustainable Development in Europe*. Environmental Issues Series 16.

Evans, R. (1996a) *Soil Erosion and its Impacts in England and Wales*. Friends of the Earth, London.

Evans, R. (1996b) Some soil factors influencing water erosion of arable land. *Progress in Physical Geography* **20**, 205–215.

Evans, R. and Cook, S. (1986) Soil erosion in Britain. *SEESOIL* **3**, 28–59.

FAO (2001) http://www.fao.org/

Gibson, C.E. (1997) The dynamics of phosphorus in freshwater and marine environments. In *Phosphorus Losses from Soil to Water* (Eds H. Tunney, O.T. Carton, P.C. Brookes and A.E. Johnston). CAB International, Wallingford, UK, pp. 119–135.

Goudie, A. (2000) *The Human Impact on the Natural Environment*. 5th Edition. Blackwell, Oxford.

Grieve, I. and Hipkin, J.A. (1996) Soil erosion and sustainability. In: *Soils, Sustainability and the Natural Heritage* (Eds A.G. Taylor, J.E. Gordon and M.B. Usher). Scottish Natural Heritage, Edinburgh, pp. 236–248.

Harrod, T. (1994) Runoff, soil erosion and pesticide pollution in Cornwall. In: *Conserving Soil Resources: European Perspectives* (Ed. R.J. Rickson). CAB International, Wallingford, pp. 105–115.

Hasholt, B. (1991) Influence of erosion on the transport of suspended sediment and phosphorus. *International Association of Hydrological Sciences Publication* **203**, 329–338.

Heathwaite, A.L. (1997) Sources and pathways of phosphorus loss from agriculture. In: *Phosphorus Loss from Soil to Water* (Eds H. Tunney, O.T. Carton, P.C. Brookes and A.E. Johnston). CAB International, Wallingford, pp. 205–224.

Heaney, S.I., Foy, R.H., Kennedy, G.J.A., Crozier, W.W. and O'Connor, W.C.K. (2001) Impacts of agriculture on agriculture systems: lessons learnt and new unknowns in Northern Ireland. *Marine and Freshwater Research* **52**, 151–163.

Hudson, N. (1995) *Soil Conservation*. Batsford, London.

Kronvang, B. (1990) Sediment-associated phosphorus transport from two intensively farmed catchment areas. In: *Soil Erosion on Agricultural Land* (Eds J. Boardman, I.D.L. Foster and J.A. Dearing). Wiley, Chichester, UK, pp. 313–330.

Leeks, G.J.L. (1995) *Effects of agricultural soil erosion on watercourses*. National Rivers Authority R&D Note 409.

MAFF (1998) *Code of Good Agricultural Practice for the Protection of Soil*. Ministry of Agriculture, Fisheries and Food, London.

MAFF (1999a) *Controlling soil erosion: a manual for the assessment and management of agricultural land at risk of water erosion in lowland England*. Ministry of Agriculture, Fisheries and Food, London.

MAFF (1999b) *Controlling soil erosion – an advisory booklet for the management of agricultural land*. Ministry of Agriculture, Fisheries and Food, London.

MAFF (2000) *Towards sustainable agriculture – a pilot set of indicators*. Ministry of Agriculture, Fisheries and Food, London.

Massa, F., Bagliniere, J.L., Prunet, P. and Grimaldi, C. (2000) Egg-to-fry survival of brown trout (*Salmo trutta*) and chemical environment in the redd. *Cybium* **24**, 129–140.

Matthiessen, P., Allchin, C., Williams, R.J., Bird, S.C., Brooke and Glendinning, P.J. (1992) The translocation of some herbicides between soil and water in a small catchment. *Journal of the Institution of Water and Environmental Management* **6**, 496–504.

Morgan, R.P.C. (1995) *Soil Erosion and Conservation*. 2nd Edition. Longmans, Harlow.

Napier, T. (1990) The evolution of US soil conservation policy: from voluntary adoption to coercion. In: *Soil Erosion on Agricultural Land* (Eds J. Boardman, I.D.L. Foster and J.A. Dearing). Wiley, Chichester, pp. 627–644.

NRA (1992) *The influence of agriculture on the quality of natural waters in England and Wales*. National Rivers Authority, Bristol. NRA Water Quality Series No. 6.

NRCS (1997) *Water Quality and Agriculture: Status, Conditions, and Trends*. Working Paper No. 16, Washington DC.

Parkinson, R.J. (1995) Soil Management. In: *The Agricultural Notebook* (Ed. R.J. Soffe). Blackwell, Oxford, pp. 87–117.

RCEP (1979) *Agriculture and Pollution*. Royal Commission on Environmental Pollution 7th report (Cm 7644). HMSO, London.

RCEP (1984) *Tackling Pollution – Experience and Prospects*. Royal Commission on Environmental Pollution 10th report. HMSO, London.

RCEP (1996) *Sustainable Use of Soil*. Royal Commission on Environmental Pollution 19th Report (Cm 3165). HMSO, London.

Reganold, J.P., Elliot, L.F. and Unger, Y.L. (1987) Long-term effects of organic and conventional farming on soil erosion. *Nature* **330**, 370–372.

Sharpley, A. and Smith, S.J. (1990). Phosphorus transport in agricultural runoff: the role of soil erosion. In: *Soil Erosion on Agricultural Land* (Eds J. Boardman, I.D.L. Foster and J.A. Dearing). Wiley, Chichester, pp. 350–366.

Skinner, R.J. and Chambers, B.J. (1996) A survey to assess the extent of soil water erosion in lowland England and Wales. *Soil Use and Management* **12**, 214–220.

Soulsby, C., Youngson, A.F., Moir, H.J. and Malcolm, I.A. (2001). Fine sediment influence on salmonid spawning habitat in a lowland agricultural stream: a preliminary assessment. *Science of the Total Environment* **265**, 295–307.

Tivy, J. (1990) *Agricultural Ecology*. Longmans, London.

USDA (2001) http://www.nhq.nrcs.usda.gov/land/env/soil1

Verstraeten, G. amd Poesen, J. (1999) The nature of small-scale flooding, muddy floods and retention pond sedimentation in central Belgium. *Geomorphology* **29**, 275–292.

Webb, J., Loveland, P.J., Chambers, B.J., Mitchell, R. and Garwood, T. (2001) The impact of modern farming practices on soil fertility and quality in England and Wales. *Journal of Agricultural Science (Camridge)*, **137**, 127–138.

Young, C.E. and Osborn, C.T. (1990) *The Conservation Reserve Program: and economic assessment.* US Department of Agriculture, Washington.

Organic Wastes 5

5.1 INTRODUCTION

This chapter deals with two distinct forms of organic waste that can cause agricultural pollution. Firstly, there are those organic wastes that are produced by agricultural activity. These are principally the excretion products (i.e. faeces and urine) of farm animals, but may also include other materials such as silage effluent and dirty water from milking parlours. Secondly, there are wastes such as sewage sludge, paper pulps and food processing wastes which, although produced off-farm (i.e. mainly from domestic household and industrial sources), are commonly brought onto farms and applied or recycled to agricultural land. These agricultural and non-agricultural organic wastes contain a range of materials including carbohydrates, fats, proteins, nitrates, phosphates and ammonia, as well as potentially being subject to contamination by pesticides, oils, veterinary products, trace metals and pathogens.

This chapter will address:

- the production and form of agricultural organic wastes;
- the pollution risks associated with organic farm wastes and non-agricultural organic wastes;
- management options and practical sustainable solutions to limit these risks.

Organic wastes and agriculture

The application of organic wastes, notably livestock manures, to land has historically been important for maintaining soil fertility in terms of both nutrient status and organic matter levels. However, with the increased specialisation and intensification of livestock farming (Chapter 1) a number of important trends have emerged. Firstly, livestock manures and slurries are produced in considerable quantities and, unlike most domestic sewage, are usually disposed of without treatment. It is estimated that the annual livestock waste production on UK farms totals some 191 million tonnes per year (Smith and Chambers, 1995). Of this approximately 80 million tonnes is slurry or manure requiring handling, storage and spreading, whilst the majority is voided directly to the land by grazing animals (Smith *et al.*, 2000). An estimated 95% of all cattle manure is returned to land in the UK, with rates of application between 30 and 300 t ha^{-1} (Johnes *et al.,* 1996; Jackson and Smith, 1997). Secondly, intensification within the livestock industry leads to large numbers of animals being concentrated in relatively small areas and consequently the production of large volumes of waste at a single site. For example, a dairy cow typically produces 50–60 litres of excreta per day, and uses upto 3 kg of straw for bedding (loose housing) and up to 35 l of water to wash the milking parlour and dairy equipment. A typical dairy farm might have 100 milking cows. Over a six-month winter period alone these would generate over 1,000,000 l of urine and faeces, plus 55 tonnes of soiled straw bedding and 640,000 l of dirty water; all of which has to be disposed of safely. Similar calculations can be made for other types of livestock (Table 5.1). Thirdly, The predominance of livestock farming in the west of the UK and arable farms in the east, means that the traditional use of farm wastes on arable land is increasingly restricted.

Many farmers have come to regard the application of livestock manures to agricultural land as little more than a waste disposal exercise (MAFF, 2000), yet the potential financial value of the nutrient content of livestock wastes is very high. It is estimated that the livestock excreta produced annually in the UK has a potential fertiliser value of over £200 million per year (Table 5.2), approximately one-third the value of equivalent nutrients applied as inorganic fertiliser (HMSO, 1993; Smith and Chambers, 1995). A single dairy cow voids approximately 21 kg of phosphate, 52 kg of potassium and 47 kg of available nitrogen in faeces and urine during a 6-month housing period.

Although organic wastes are a valuable alternative to artificial inorganic fertilisers, some farmers have relatively little perception of their nutrient value and consequently may manage them inefficiently, although many claim to make allowances for the nutrients supplied in manures in fertiliser plans (Owen, 1998; MAFF, 2000; Smith *et al.*, 2001). This may lead to:

- Poor utilisation of their nutrient content (Chapter 2): For example, farmers may make only small and/or inconsistent reductions in the use of fertilisers following the application of livestock slurries and manures, whereas they could be making significant savings in fertiliser costs without loss of crop yield (Smith *et al.*, 2001).
- Risk of environmental pollution: A common practice has been the application of large amounts of slurry and manure to relatively small areas of land during the autumn and winter months when there is a risk of rainfall causing surface run-off and/or nutrient leaching.

Table 5.1 Examples of amounts of excreta produced by livestock in respect to body weight (MAFF, 1998a).

		Excreta	
Type of livestock	Body weight (kg)	Moisture content (%)	Typical volume (litres day^{-1})
1 dairy cow	550	90	57
1 beef bullock	350	90	27
1 pig, dry meal fed	60	90	4
1 pig, whey fed	60	97	14
1000 laying hens	2000	70	115

The polluting effects of organic wastes can occur in a number of ways and therefore require a broad range of approaches to control. All watercourses (notably lakes, ponds, rivers, streams and field ditches) adjacent to areas of production, storage or application of organic wastes are potentially at risk of point source pollution. This risk is different from the diffuse pollution that occurs when the microbial breakdown of manure applied to the soil occurs out of phase with the N uptake of a growing crop and leads to nitrate leaching.

The remainder of this chapter describes the nature and causes of pollution by agricultural and non-agricultural organic wastes on farms, and introduces some practical solutions to minimise environmental risks. The pollution problems addressed in this chapter are principally point source; the threat of the diffuse pollution of ground and surface waters by N and P from organic wastes has already been discussed in Chapters 2 and 3. Gaseous emissions and odour nuisance are addressed in Chapter 6.

Table 5.2 Potential value in millions of pounds, of organic manures produced in the UK compared with the cost (shown in parentheses) of equivalent nutrients supplied as inorganic fertiliser (Dampney, 1995).

	Available N	Total P	Total K	All nutrients
Cattle	29	34	58	121
Pig	14	12	7	33
Poultry	13	25	12	50
Total value	56 (410)	71 (100)	77 (80)	204 (590)

Organic wastes cannot only cause pollution within the wider environment, but may also pose problems for crops and livestock on the farm if care is not taken to utilise or dispose of them properly (Table 5.3). Crop smothering and scorch are commonly associated with the application of organic manures and slurries (notably to grassland), and can lead to significant yield reductions due to physical crop damage and reductions in nutrient uptake (Smith and Chambers, 1993). Nutrient imbalances following organic amendments to soils may also cause temporary macronutrient deficiencies, reducing seedling emergence and restricting plant growth (Tisdale *et al.,* 1993; Glendinning, 1999; Van Kessel *et al.,* 2000). Sewage sludge application can also cause problems on the farm due to its content of potentially toxic elements such as heavy metals, organic micropollutants and microbial pathogenic organisms (Smith, 1996). These problems are considered in more detail later in this chapter.

Table 5.3 Typical problems caused through organic waste application for crops and livestock.

Winter kill	Application of organic wastes to crops during the winter may smother small seedlings, as well as promote growth during mild weather thereby increasing the risk of winter kill
Scorch	Damage can be caused by direct contact between slurry and the crop
Fouling	Livestock may not feed on grassland tainted with organic waste
Disease	Diseases (parasites, bacteria, etc.) may be passed onto livestock
Staggers	Heavy application can cause nutrient imbalance in growing herbage and subsequently on grazing livestock (e.g. magnesium/potassium imbalance, known as hypomagnesaemia)
Flies	Excessive application can cause environmental nuisance

5.2 FARM WASTES

Four main classes of farm waste can be identified:

Slurries The Control of Pollution (Silage, Slurry and Agricultural Fuel Oil) Regulations (HMSO, 1991) define slurries as 'excreta produced by livestock whilst in a yard or building, or . . . a mixture consisting wholly or mainly of such excreta, bedding, rainwater and washings from a building or yard used by livestock or any combination of these; of a consistency that allows it to be pumped or discharged by gravity at any stage in the handling process'. Or alternatively, a slurry could be considered as being a mixture faeces, urine and water, but has less than 10% dry matter (Shepherd and Gibbs, 2001);

Solid Manure This includes waste materials with a high solids content, such as straw from deep litter or traditional covered yards, which is usually present in sufficient quantity to allow the manure to be stacked. Solid manure is usually handled as a solid, usually >10% dm, but when fresh or recently stacked can be very strawy and bulky with a low density (<0.5 tm^{-3}). Old manure, which has been stacked for a period of months, will have been subject to microbiological decomposition (composting if aerobic) and gradually becomes darker and more friable when compared with fresh manure;

Dirty Water This includes water used to wash down yards, milk parlours, farm buildings, contaminated rainwater, etc. and is defined by MAFF as 'a waste, generally less than 3% DM made up of water contaminated by manure, crop seepage, milk or other dairy products or cleaning materials';

Silage Effluent This includes the effluent from a range of forage crops, principally grass, when enclosed in a pit, silo or big bale during the process of making silage.

Changes in farming systems

Livestock farming has changed considerably since the late 1940s with the advent of more specialised and intensive production systems. With this intensification have come a number of manure and slurry management problems for farmers. For example, there has been a steady increase in the number of dairy cows kept in the UK with the result that the average herd size increased from 18 cows in 1955 to 63 cows in 1991 (HMSO, 1993). Herds can range in size from less than 30 to greater than 400, although the larger herds tend to be subdivided into groups of 100–150. Furthermore, between 1930 and 1990, livestock density increased in the UK from 0.83 head ha^{-1} to 1.77 head ha^{-1} (Edwards and Withers, 1998). With the trend towards larger, more intensively-managed herds, comes the problem of much greater volumes of waste to manage. This is caused not only by the increase in herd size, but also by changes in management, such as the indoor housing of livestock for a greater proportion of the year. A few herds are kept indoors all year round (so-called 'zero grazing').

The type of wastes produced depends largely upon how the animals are housed. Dairy cows, for instance, can be housed in one of two ways:

Deep litter In this system, the housed cows have access to a covered area of 'deep litter' bedding material (usually straw) where they can lie and sleep. As more clean litter is regularly added to the area, the waste bedding accumulates to the point that it is necessary to clear it out. The resultant solid waste ('farm yard manure' or FYM) is a mixture of straw, urine and faeces (Figure 5.1).

Cubicles The cows are housed indoors in large specialist buildings. Individuals are usually housed in cubicles with a minimum of bedding material, such as sawdust or chopped straw, and a mechanised cleaning system (e.g. a tractor and yard scraper) to collect slurry from a central gutter (Culpin, 1992).

Similar distinctions in housing type exist for other forms of livestock, but the general rule is that deep litter systems tend to produce solid waste, whilst specialist livestock 'houses' produce slurries (almost 45% of the 10.3 million tonnes of manure from pig production in the UK is as slurry) (Smith *et al.*, 2000). An exception to this would be the poultry sector in which the 70% of manure from broiler and egg production is litter-based with sawdust and wood shavings as bedding material (Smith *et al.*, 2001).

Figure 5.1 Deep litter housing system.

Livestock that are housed require a source of food, and for over-wintering cattle (beef and dairy) this is usually provided in the form of grass conserved as hay or silage. Hay-making was traditionally the most popular method of conservation, but has now been largely replaced by silage. This effectively involves the 'pickling' of grass (and other forage crops) in the lactic acid which is produced naturally by the fermentation of water-soluble sugars in a crop when it is cut, chopped and heaped into an airtight silo or sealed into a big bale.

An undesirable product of silage-making is silage effluent, a mixture of water, soluble sugars and organic acids which begins to flow from the silo within a day or so of filling, and can continue for several weeks. This is one of the most environmentally degrading organic waste materials on the farm, since it is both corrosive and extremely polluting (Lennox *et al.*, 1998; Richardson *et al.*, 1999). Selected silage effluent characteristics are given in Table 5.4 and highlight the potentially corrosive and damaging effect it may have on and off the farm (O'Donnell *et al.*, 1995).

Pollution risks from farm wastes

Dissolved oxygen is continually consumed by the aquatic organisms found in watercourses, but is usually replenished by a number of natural processes. These include: re-aeration due to the physical reaction of air and water (e.g. turbulence); photosynthesis by water-borne plants and algae; decreases in temperature which lead to a reduction in micro-organism activity and an increase in the oxygen saturation potential of the water; dilution by more highly oxygenated water from other sources (Nemerow, 1991). However, this balance can be upset if an organic pollutant enters a watercourse. Characteristically, this acts as a substrate or food for a number of aquatic micro-organisms (notably bacteria and protozoa) and macro-invertebrates stimulating a huge increase in their population and consequent consumption of dissolved oxygen (Figure 3.7). If the levels of organic substrate are sufficiently high, populations of micro-organisms (often evident as a pale carpet of sewage fungus a host of micro-organisms dominated by *Sphaerotilus natans* on the watercourse-bed) may build-up to the point at which they consume oxygen more rapidly than it can be replenished, thus leading to a net depletion of oxygen (Mason, 1996). This can have a serious impact upon the population of a number of 'clean water' organisms. Fish have a relatively low tolerance to oxygen depletion (Table 5.5) and can suffer significant losses for large distances downstream of a serious pollution incident.

Table 5.4 Selected characteristics of silage effluent, collected from unwilted grass silage (O'Donnell *et al.,* 1995).

Characteristic	
pH	3.8
Titratable acidity (mmol NaOH l^{-1})	177
Lactic acid (g kg^{-1})	24
Acetic acid (g kg^{-1})	3.3
Volatile fatty acids (acetic, propionic and butyric acid) (g kg^{-1})	3.5

A commonly used measure of the 'relative pollution potential' of an organic contaminant is the amount (mg l^{-1}) of oxygen needed by micro-organisms to break down the material. This is called the Biochemical Oxygen Demand (BOD). It is calculated by a standard test involving the incubation of a sample of organic material in the dark over a fixed period (5 days) and at a constant temperature (20 °C). Levels of oxygen (mg l^{-1}) in the sample are measured before and after incubation, and the difference is recorded as the BOD. High levels of BOD indicate the presence of a potentially serious pollutant that should not be released into a watercourse. All organic farm wastes fall into this category (Table 5.6), with many serious pollution incidents occurring due to the poor containment or disposal of slurry and silage effluent. Silage effluent has a very high BOD and so it is important to avoid the release of even very small quantities into watercourses (Table 5.6).

The pollution problems associated with silage effluent are exacerbated by two further factors, firstly, the effluent is highly corrosive (Table 5.4) and can easily escape through silo floors, collection channels or storage tanks which are damaged, corroded, cracked or porous (Richardson *et al.,* 1999). Secondly, depending upon the moisture content of the forage being conserved, the volume of effluent produced during silage making can be very large.

Table 5.5 Relative tolerance to oxygen depletion by some river organisms.

Common name	Species	Tolerance
Salmon	*Salmo salar*	Low
Brown Trout	Salmo trutta	Low
Shrimp	*Gammarus pulex*	Medium
Water Hog-Louse	*Asellus aquaticus*	Medium
Chironomid midge	*Chironomus riparius*	High
Blood-worm	*Tubifex tubifex*	High

Effluent production from grass silage in a horizontal clamp is only insignificant when the grass has been wilted to less than 65% moisture content. Lennox *et al.* (1998) attributed the 51% decline in agricultural pollution incidents in Northern Ireland between 1987 and 1995, in part to a 78% decline in silage effluent pollution due to an observed increase in grass wilting and favourable weather in the summer months. At the recommended moisture content of 75%, up to 110 l of effluent per tonne of silage can be expected (i.e. 55,000 l from a 500 tonne clamp). While fresh, unwilted grass at about 85% moisture content may yield up to 330 l of effluent $tonne^{-1}$ of silage (i.e. 165,000 l from a 500 tonne clamp) (MAFF, 1993).

Table 5.6 The Biochemical Oxygen Demand (BOD, mg l^{-1}) of farm wastes in comparison with other organic materials (NRA, 1992).

Organic material	Typical BOD
Clean river water	<5
Untreated human sewage	350
Yard washings	2,000
Animal slurry	30,000
Silage effluent	60,000

The specific effects arising from an organic waste entering a watercourse can be difficult to predict since they depend upon many factors, including temperature, dilution rates and the type of watercourse (Box 5.1). Additionally, time of year is important – silage effluent production in early summer is often more of a problem than winter produced materials (O'Donnell *et al.,* 1997). However, in time, watercourses are also able to recover from organic pollution by a process of self-purification, whereby the organic material is broken down and oxygen levels are replenished by natural processes (Mason, 1996).

Ammonia in watercourses

The relatively high levels of ammonia found in some organic wastes may pose an additional pollution threat to fish and to some freshwater invertebrates from livestock wastes entering aquatic systems (Hickey *et al.,* 1999). Importantly, the threat is not only to surface waters but also groundwater resources (Stone *et al.,* 1998; Arnold and Meister, 1999; DETR, 2001). In a study of the impact of dairy farming on river quality in south-west Wales Schofield *et al.* (1993) observed the expected water quality problems associated with high BOD, but also noted high ammonia levels as a significant aquatic pollutant. In one stream, background levels of ammonia of between 3 and 5 mg N l^{-1} were found with peak levels as high as 20 mg N l^{-1}; levels which were sufficiently toxic to prevent the maintenance of a sustainable fishery (Tables 5.7 and 5.8). The study concluded that tributaries in the catchment which had high ammoniacal-N and BOD were those which drained the areas of land used for dairy farming. Furthermore, it was the production of manures and slurries, and washing of yards and milking parlours that was causing a measurable deterioration in the quality of watercourses. Reductions in water quality during the study were also related to rainfall. Periods of heavy rainfall caused the increased run-off of slurry from yards as well as from fields that had recently received slurry applications.

An increase in free ammonia in watercourses may also inhibit the process of nitrification in sediments and so cause the potentially toxic build up of nitrite in the water body (Chapter 2) (Smith *et al.*, 1997).

Box 5.1 UK case studies of pollution incidents caused by farm waste

The Environment Agency is responsible for monitoring the water quality of all rivers, lakes and streams. In addition to routine monitoring, surveys are conducted when pollution incidents occur and this information may be used in prosecutions. The incidents described below all occurred in the north-west of England.

Silage effluent

A small river was contaminated with silage effluent from a nearby farm and caused severe pollution. The pollution source was traced to a drain; downstream of the drain, the riverbed was coated with growths of the aquatic fungus *Leptomitis lacteus* whilst upstream the riverbed was clear. A survey of the biology of the stream showed that there were 16 families of freshwater invertebrates (insects, snails, shrimps, etc.) above the drain whilst there were only three downstream for a distance of at least 500 m.

On-farm food processing waste

This incident was caused by organic wastes from a farm dairy specialising in yoghurt making. The effluent entered a stream via a tributary; upstream of the confluence the stream was clean and 11 families of invertebrates were recorded. Downstream of the confluence no invertebrates were found, whilst the stream-bed was coated with sulphur bacteria which are indicative of anoxic (oxygen-free) conditions. Further downstream, thick growths of 'sewage fungus' occurred (predominantly *Sphaerotilus natans*). Samples taken during the investigation showed the effect on the chemistry of the stream:

	Dissolved oxygen (%)	Suspended solids (mg l^{-1})	BOD (mg l^{-1})	Conductivity (μS cm^{-1})	Ammonia (mg N l^{-1})
Above tributary	74	9	5.4	578	1.0
Tributary	1	36	146	839	12.2
Below tributary	59	11	10.3	608	1.9

Slurry
This incident was reported by the farmer who had lost a quantity of slurry into a stream after a storage tank had leaked. Downstream of the point where slurry entered the stream, the only invertebrates living were red chironomids and tubificid worms – both are very tolerant of pollution. All the invertebrates indicative of clean water were dead. It was concluded that the stream had been polluted previously and that this incident caused a worsening of an already chronic problem. Chemical sampling also showed the effect of this organic pollution:

	Dissolved oxygen (%)	Suspended solids (mg l^{-1})	BOD (mg l^{-1})	Conductivity (μS cm^{-1})	Ammonia (mg N l^{-1})
Above pollution	83	6	2	282	0.01
Below pollution	18	255	557	1276	45.8

Pathogens from farm wastes

Farm livestock are potential carriers of a number of pathogenic bacteria, viruses and parasites. These can be transferred to humans if the application of animal waste to land contaminates crops or watercourses (DoE/DH, 1990). In Germany, the bacterial contamination of a crop of parsley with *Citrobacter freundii* following the application of pig slurry to a private garden caused an outbreak of severe gastroenteritis and associated syndromes amongst children at a nursery school (Tschäpe *et al.,* 1995). Nine children were hospitalised, including one who later died. Subsequent analysis of the fresh parsley involved found it to be also highly contaminated with many other bacterial species, including *Proteus mirablis Pseudomonas* spp. and *Escherichia coli.*

Cattle are the primary reservoir of the virulent human pathogen *E. coli* serotype O157, which causes in excess of 1000 cases of food poisoning in the UK each year. Whilst cattle show little observable effects from *E. coli* O157 (up to 15% of UK cattle herds may be infected) human infection may be very serious with a mortality rate of about 5%. Exposure to humans may occur through the contamination of food by faeces, manure, slurry and even through the inadvertent consumption of contaminated soil. *E. coli* may remain viable in non-aerated manures for almost 12 months, but in aerated manure piles and soils this is often reduced to between 2 and 4 months and just 10 days in cattle slurries (Jones, 1999).

A further specific problem identified in the UK by MAFF (1998a) is the protozoan parasite, *Cryptosporidium,* found in many animal wastes. This can contaminate water courses via leaching/run-off and lead to the risk of human infection if water is subsequently extracted for drinking supplies (*Cryptosporidium* is resistant to many of the procedures used to treat drinking water). In the UK, the incidences of cryptosporidiosis rose nearly 10-fold in cattle and 5-fold in sheep between 1983 and 1994 and it has been postulated that *Cryptosporidium* is now ubiquitous amongst mammals in the UK (Nicholson *et al.,* 2000).

Veterinary products in farm wastes

According to the RCEP, the contamination of livestock wastes with certain veterinary products, or their metabolites, can have unwelcome environmental effects (RCEP, 1996).

One example is the anti-parasitic drug, ivermectin. This is an effective and popular alternative to the OPs (Chapter 7) commonly used in modern livestock husbandry, but it is exceptionally persistent and residues in the manure of treated animals can significantly reduce the number of variety of insects found in and around manure. There is concern that this might in turn affect insect-eating birds and mammals (particularly bats).

Table 5.7 Analysis of chemical composition of farmyard slurry at four farms (at 4 dates) in the Eastern Cleddau catchment, Dyfed, Wales (Schofield *et al.*, 1993).

Farm and Date	BOD ($mg l^{-1}$)	Ammoniacal N ($mg N l^{-1}$)	Total solids (%)	Total K (%)	Total P (%)	Total N (%)
1 16/3/88	–	–	9.0	2.6	0.53	1.92
2 5/5/88	12,400	850	6.2	–	0.70	2.30
3 30/5/89	5,600	1,320	9.1	–	0.77	2.16
4 7/7/88	13,600	–	7.9	2.6	0.51	1.98

Water pollution incidents

The intensification of livestock production in the UK has greatly increased the risk and occurrence of water pollution by organic wastes. Organic pollutants account for almost 90% of all farm pollution incidents recorded, with dairy farming causing more incidents than all other agricultural sources combined. Indeed, 45% of Category 1 incidents were attributed to dairying in 1999 (Environment Agency, 2001a). The category of pollution incident ranges from 1 to 3, with 1 being the most severe in regard to ecological and environmental damage (Chapter 7). Not surprisingly incidents are concentrated in areas of high livestock production, and so in the UK primarily occur in the south west, midlands and north west of the country (Figure 5.2) (DETR, 2001). A significant number of incidents involving pig slurry and poultry manure also occur in East Anglia where intensive units complement arable farming enterprises. Problems most commonly arise from acute point source pollution due to:

- the poor containment of manures, slurries and silage effluent e.g. run-off from dirty yards, leaking drains or the structural collapse of slurry stores;
- the discharge of dairy washings. As well as the high BOD of milk, dairy washings also pose a hazard because of the hypochlorite-based solutions used to sterilise milking equipment which if discharged to a watercourse can significantly increase its chlorine content (Conway and Pretty, 1991);
- Surface run-off of slurry following application to land. This mainly occurs during periods of heavy rainfall when the soil is waterlogged or frozen, and is especially pronounced when the land is sloping (NRA, 1992; MAFF, 1998a).

Foy and Kirk (1995) carried out a regional study of water quality and the influence of agriculture in Northern Ireland by monitoring 21 streams in two river catchments. Water quality was measured on a fisheries ecosystem scale (Table 5.8) of FE1 (good, salmonoid water) to FE6 (bad, fish absent). The study concluded that water quality measured on this scale was inversely correlated with the stocking rate of grazing animals and that a decrease in water quality was associated with an increase in the combined grazing/stocking rate of cattle and sheep. Stocking rates were positively correlated with maximum BOD, total ammoniacal nitrogen and minimum dissolved oxygen and are all indicators of poor water quality. An earlier study in Northern Ireland also identified the increased use of silage as a feed for livestock as a cause of greater pollution incidence (Foy *et al.*, 1994). A reduction in water pollution incidents in the same area was attributed to a reduction silage production (Lennox *et al.*, 1998).

From the mid-1980s, all UK regional water authorities undertook farm visit campaigns (continued by the National Rivers Authority and the Environment Agency) to identify sources of pollution and to provide advice on remedial action. Despite some concerns that it would take many years for good practice and improved storage technologies to significantly improve water quality (NRA, 1992), these campaigns appear to have been partially successful since the number of recorded farm pollution incidents in the UK fell by 12% between 1985 and 1993 (NRA, 1995). However, it appears that an increase in financial penalties for causing pollution, grant aid and targeted advice in catchments given by agriculture advisory staff have still to make a really significant positive impact in reducing agricultural water pollution incidents (Lennox *et al.*, 1998).

5.3 NON-AGRICULTURAL ORGANIC SOIL AMENDMENTS

Sewage sludge or biosolids

Sewage sludge (biosolids) is the residual material from sewage plants treating domestic or urban waste water (MAFF, 1993; Renner, 2000). It is a major waste product of the water industry being produced in increasing quantities as additional sewage treatment is required to reduce the polluting impact of discharges from waste water systems (RCEP, 1996). The total amount of sewage sludge produced per year in the UK is estimated to be 1.1 million tonnes (Environment Agency, 2001a). With more stringent clean water regulations taking effect over much of Europe it is likely that this will increase in the future (Renner, 2000), with an estimated 50% increase in sludge production across Europe by 2005 (CEN, 1999; EU, 2000). Currently, sewage sludge makes up less than 4% of the total organic wastes produced on a dry solids basis across Europe (livestock wastes contribute about 90%). Importantly sewage sludge quality, use and monitoring requirements are prescribed in great detail – it is the most controlled of all organic wastes (CEN, 1999).

Spreading and injection to plough depth to agricultural land is currently the most common destination for sewage sludge (Figure 5.3). Yet only 0.5% of the total area of agricultural land in the UK receives sludge, and of this 59% is arable. This compares with almost 16 million tonnes of farm manure (dry solids) and 4 million tonnes of industrial wastes (fresh weight) that are recycled to land (Hickman *et al.*, 1999).

Table 5.8 Classification of water quality based on suitability as fishery for salmonoids or cyprinids (DoE, 1993).

Class	Description	DO (%)	BOD (mg l^{-1})	Ammonia (mg N l^{-1})
	Percentile	10	90	90
FE1	High class salmonoid	80	2.5	0.25
FE2	Sustainable salmonoid	70	4.0	0.6
FE3	High class cyprinid	60	6.0	1.3
FE4	Sustainable cyprinid	50	8.0	2.5
FE5	Fishery not sustainable	20	15.0	9.0
FE6	Fishery not sustainable	–	–	–

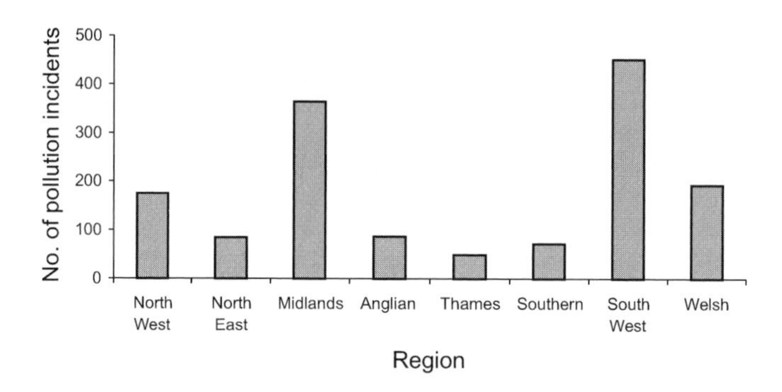

Figure 5.2 Distribution of total pollution incidents from organic waste in 1999, not including sewage sludge (DETR, 2001).

Sewage sludge is a complex organic material derived mainly from human wastes. It may be supplied free to farmers by water companies and can have a number of potentially beneficial effects when applied to agricultural land:

- All sludges contain significant quantities of N, P (Table 5.9), and micronutrients such as essential trace metals and it is useful to recycle as much of this back to agricultural land as possible. The potential savings in fertiliser applications by farmers who use sewage sludge have been estimated to be over £15 million per year (Hall, 1992).

- Sewage sludge can also act as a soil improver by adding considerable amounts of organic matter, thus enhancing the structure and water retention capacity of some soils, especially when added as a drier processed sewage sludge product called cake (White, 1997).

- Sludge can also be treated to increase its agricultural value and disposal potential, for example by adding cement kiln dust to impove its liming properties or composting with green wastes (Rund, 1995; Barry *et al.,* 1998; CEN, 1999; Chaney *et al.,* 2000).

The full value of sewage sludge as a nutrient source and soil improver is, however, constrained by its contamination with a number of potentially harmful substances, notably heavy metals and organic micropollutants, as well as pathogenic micro-organisms such as *Salmonella* spp., *Taenia saginata*, *Asaris* and *E. coli* O157:H7 (Carrington *et al.*, 1998).

Table 5.9 Typical nitrogen and phosphate content of sewage sludges (Scottish Envirotec, 1995).

		Nitrogen		Phosphate (P_2O_5)	
	Dry solids (%)	Total (%)	% Available	Total (%)	% Available
Liquid digested	4	0.20	60	0.15	50
Digested cake	25	0.75	15	0.90	50
Dried sludge	94	3.10	15	4.12	50

Pollution risks from applying sewage sludge to agricultural land

Sewage sludge is not just derived from domestic sewage and wastewater, but may also receive inputs from industrial effluents and road run-off. It usually contains a number of metals including cadmium (Cd), copper (Cu), chromium (Cr), nickel (Ni), lead (Pb) and zinc (Zn), as well as organic micropollutants such as polynuclear aromatic hydrocarbons (PAHs) and polychlorinated biphenyls (PCBs) (Smith, 1996). Importantly, it is thought that the organic contaminants present within sewage sludge pose minimal risk to the wider environment and humans (Carrington *et al.*, 1998), although limit values for organic micro-pollutants in sludges for use on land are likely to be set by the EU (EU, 2000). Concentrations of contaminants in sewage sludges vary greatly according to the source of the sludge and the nature of the catchment area. In industrialised areas, point source contamination by trace metals may include Cr and Ni released by the iron and steel industry, Cd and Pb from the manufacture of batteries, and Zn from zinc plating works.

However, with greater controls imposed by water companies over industrial discharges to sewers (EU, 2000), plus a general decline in manufacturing industry, the contamination of sewage with industrial effluents has fallen significantly in recent years (Environment Agency, 2001b) (Figure 5.4). However, considerable quantities of contaminants may still enter sewers from a number of more diverse and diffuse sources that are difficult to control, including:

- dichlorobenzene from toilet cleaner and alkyl benzenes from detergents;
- Cu and Zn from domestic products such as shampoos, skin creams, toilet cleaners and mouth wash;
- Cu and Zn from plumbing fittings, water pipes and storage tanks;
- Hg from dental surgeries.

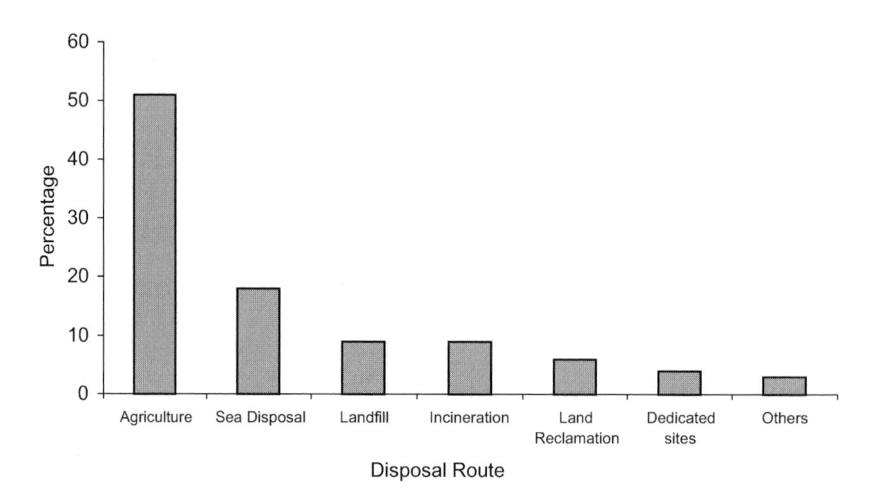

Figure 5.3 The fate of sewage sludge in the UK in 1996/7 (Others includes forestry, horticultural compost and disposal within the confines of the treatment works) (Environmental Agency, 2001b). Sea disposal was phased out in 1998 enforced through EC Directive EC91/271/EEC).

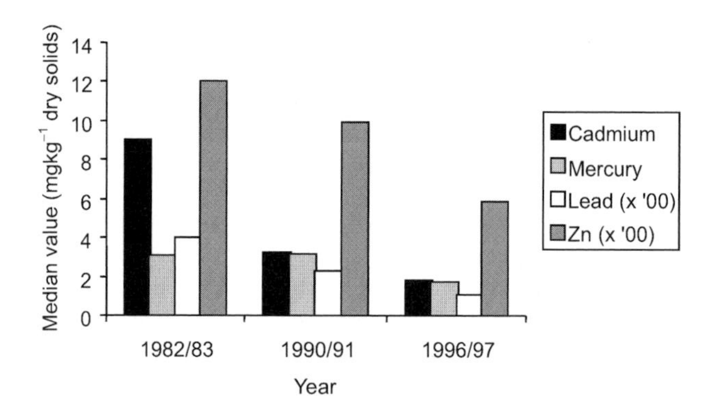

Figure 5.4 The change in selected metal content of sewage sludges applied to agricultural land in the UK (Environment Agency, 2001b).

Although trace metals are a natural component of the environment and some are essential in small amounts for healthy plant and animal life, there is concern that they are present in agricultural systems in increasingly large amounts. Sewage sludge is one of the most important sources of trace metal inputs on farms, but by no means the only one (Box 5.2). The primary concern is that if sewage sludges containing potentially toxic elements are applied too frequently to agricultural land, or over a prolonged period, metals, in particular, are liable to accumulate and be retained in the cultivated layers (Box 5.3) (Smith, 1996). However, soil characteristics such as pH, organic matter content, hydrous oxides of iron (Fe) and manganese (Mn), redox condition and clay content will greatly influence the ultimate fate of these metals (Alloway, 1995). It is a possibility that elements such as Cd, Cu and Zn could eventually build up to levels in soils that may cause detrimental effects upon soil microbial populations, soil animals, crop plants and potentially enter the human food-chain (McBride, 1995; Renner, 2000).

The toxicity of a trace metal, and its effect upon plant growth and yield, is the product of its concentration or dose, not simply its presence (Alloway, 1995). For example, Cu and Zn are essential micronutrients and required by plants in optimal amounts; if these levels of uptake are exceeded then phytotoxicity occurs. Equally, plants can tolerate low levels of non-essential metals (i.e. those with unknown beneficial biological function), such as Cd, Pb and mercury (Hg), but suffer phytotoxicity if their uptake of these increases (Reuter and Robinson, 1997). It has been suggested that imbalance of nutrient content in herbage, caused through the application of sewage sludges, may adversely affect grazing livestock. It is thought that these may be due to antagonistic effects of a particular trace element in high concentrations in the sludge (e.g. Fe) reducing the uptake of another trace element (e.g. Cu) (Carrington *et al.*, 1998).

Box 5.2 Other sources of trace metals to agroecosystems

Potentially toxic elements are added to agricultural land in Europe from a range of sources. The relative importance of these sources, in regard to quantities and the particular elements of importance for each source, in comparison to sewage sludges can be seen in the following table. Potentially toxic element additions to agricultural holdings in England and Wales are given in tonnes per year (Carrington *et al.*, 1998; CEN, 1999).

Source	Zn	Cu	Pb	Cd	As
Sewage sludge	388	237	113	2.1	1.8
Atmospheric deposition	2543	647	485	20.8	33.5
Fertilisers	246	96	11	11.9	9.0
Lime	44	10	8	1.2	0
Agrochemicals	26	8	0	0	0
Animal manures	1992	710	52	4.1	16.2

Livestock wastes

Soil contamination may occur due to the repeated application of livestock wastes that contain metals from additives put in feed to improve conversion efficiency and accelerate growth rates. It occurs primarily with pigs and poultry manures that may contain elevated concentrations of Cu and Zn. Since the animal absorbs only a small fraction of the metal additives, the remainder is excreted as a residue in the pig slurry or poultry manure. The concentration of metals in these manures is highly variable depending upon differences in the composition of feed and pest control practices. Concentrations of Cu and Zn in pig and poultry slurries have been observed to be in the range of 300–2000 mg Cu kg^{-1} and 200–1500 mg Zn kg^{-1} (Alloway and Ayres, 1993).

The effects of elevated Cu and Zn levels in the soil remain unclear. While Cu concentrations in soils that have been regularly used for slurry disposal have increased, little uptake and translocation of Cu by crop plants is observed (Huysman *et al.*, 1994). However, it is highly likely that soil microbial populations and the essential cycling of elements, fundamental for the maintenance of soil fertility, in which microbes play a major role, may be adversely effected (Giller *et al.*, 1998).

Fertilisers, pesticides and limes

A widespread source of Cd in agricultural soils is from phosphatic fertiliser, either in the form of superphosphates (in straight P-fertilisers) or ammonium phosphate (in compound N, P, K fertilisers) (McLaughlin *et al.*, 1996). This represents 29% of the annual Cd addition to agricultural soils in the UK (Carrington *et al.*, 1998). The raw material for phosphatic fertilisers is rock phosphate and this is dissolved in sulphuric, nitric or phosphoric acid according to fertiliser material being produced. Rock phosphate (depending upon its source) may contain 5–55 mg Cd kg^{-1} (34–380 mg Cd kg^{-1}P), some 60–95% becomes incorporated into the final product depending upon the manufacturing process. Fertilisers in the Netherlands and Sweden have been found to typically contain 2–30 mg Cd kg^{-1} and a survey of 66 fertiliser samples in the UK found a mean Cd content of 37 mg kg^{-1} (Alloway, 1995). One of the most serious concerns about the steady rise of Cd levels in soils is that it is relatively mobile and readily taken up by crops (the concentration in plants may be an order of magnitude greater than in the soil in which the plant is growing), leading to the risk of increased levels in the human diet (Alloway, 1995; McLaughlin *et al.*, 2000). Cadmium may accumulate in the human body and is implicated in the causation of kidney disorders, skeletal abnormalities and also some cancers.

Agricultural limestones and pesticides may also contain metal contaminants as unwanted impurities. In a study on the metal contributions to soils of four inorganic fertilisers, two herbicides and one fungicide it was shown that all contained significant concentrations of metals, but it was primarily Zn, Co and Pb that were added (Gimeno-García *et al.*, 1996). Limes used in normal agricultural practice may also contain significant quantities of Cd (2.8–17.0 mg kg^{-1}), Cr (6.5–50 mg kg^{-1}) and Pb (15–550 mg kg^{-1}) (McLaughlin *et al.*, 2000). It is estimated that 45% of the total annual Cu and Zn additions to agricultural soils in the UK are from animal manures, 24% and 35% respectively from aerial deposition and 17% and 10% from sewage sludge (Carrington *et al.*, 1998).

Aerial deposition

The deposition of a range of metals to agricultural land from point sources has lead to localised increases in soils (MAFF, 2000). Aerial deposition is more important for some metals than others and it may account for as much as 70% of the Pb and 50% of the Cd added to agricultural soils in the UK annually (Carrington *et al.*, 1998). However, the effects tend to be very localised, depending upon local and regional industry, prevailing weather conditions and topography.

N and P pollution from sewage sludge

As with all organic wastes, poorly managed winter applications may result in leaching of nutrients into watercourses (Chapters 2 and 3). However, with sewage sludge the treatment process has a considerable influence upon nutrient content (Table 5.9) (Environment Agency, 2001b). Generally, it is liquid sludges that can cause the greatest problems, as a larger proportion of their total N is in the available ammonium form, which may subsequently be leached rapidly following nitrification (Scottish Envirotec, 1995). Injection of sludge to a depth of 10–15 cm can reduce surface run-off and place the nutrients in the most biologically active portion of the soil (Figure 5.5).

A considerable issue with the disposal or recycling of sewage sludges to agricultural land in regard to N and P pollution is the significant variability in available nutrient content of sludges from the same treatment stream over time. This can make fertiliser plans worthless in that variations of available N applied may be up to 100% greater than thought, if data from previous batches are used as reference (Stehouwer *et al.*, 2000).

Box 5.3 Woburn Market Garden Experiment

The Woburn Market Garden Experiment was set up in 1942 in order to assess the fertiliser value of various organic manures, including sewage sludge. Manures were applied annually to plots at the site from 1942 until 1961, after which no further applications were made. A total of 25 applications of sewage sludge were made to the plots at two rates approximately equivalent to 8.2 and 16.4 t^{-1} ha^{-1} $year^{-1}$. Soil samples were taken from the plots at irregular intervals (1942, 1951, 1960, 1967, 1972, 1980 and 1983) from the start of the experiment, dried and archived (McGrath and Cegarra, 1992). By the end of the sludge trial relatively large amounts of metals, particularly Zn, Cu and Ni, had accumulated in the soils and crops grown on the plots when compared to untreated control plots (LeRiche, 1968). The following table shows the range of total concentrations ($mg\,kg^{-1}$) of selected heavy metals in the treated soils in 1961 and the concentrations of metals in the same plots over 25 years later (McGrath, 1984; MAFF, 1993).

	Zn	Cd	Pb	Cu	Ni
Control	75	0.2	30	23	14
1961	340–635	8.8–19.4	131–209	125–239	30–42
1987	63–435	1.5–13	36–175	15–150	13–33

The data shows the long-term residence time of heavy metals applied to soils in sewage sludge. Over 25 years after the final application of sewage sludge to the plots there appears to be only a relatively small loss of metals, which may be attributed to the movement of soil out of the plots during cultivation.

Furthermore, 25 years after the cessation of applications of sewage sludge at this site, measurements of the soil biomass (soil microbial populations) within the treated plots revealed that it was only half that of the untreated controls (Brookes and McGrath, 1984). This was thought to be attributable to the Zn concentrations, which although below the current UK limits, may still have had a deleterious affect upon the biomass (Giller *et al.,* 1998). However, Cd concentrations in the plots are 2–3 times greater than UK limits (MAFF, 1993) and may also have contributed to the reduced microbial biomass observed on the treated plots.

This example demonstrates the long-term impact of heavy metals applied to soils in sewage sludge upon the agroecosystem. It also highlights the difficulty in developing a protective regulatory framework for the safe management of metals in sewage sludges applied to agricultural soils (Renner, 2000).

Odours

Odour nuisance can be a serious environmental problem (Chapter 6) and depends on the type of sludge treatment, method of application and prevailing weather conditions. Anaerobic or aerobic digestion of sludge often removes much of the odour nuisance, as does rapid incorporation of sludge by injection directly into the soil. However, conventional spreaders (Figure 5.6) and non-incorporation of sludge can lead to odours, as well as presenting an increased risk of surface run-off.

Pathogens

As with livestock wastes, there is a risk of pathogen contamination whenever sewage sludges are applied to agricultural land. Potentially, serious pathogens include bacteria such as *Salmonella* (especially *Salmonella typhimurium* DT104), human viruses such as Hepatitis A, parasitic nematodes and worms, and parasitic protozoa such as *Cryptosporidium*. Although the accidental contamination of soft fruit and vegetables with human faeces has been recorded as causing various infections in the UK, no link has yet been established between controlled sewage sludge application and public health problems (RCEP, 1996).

Figure 5.5 Slurry/sludge injection.

The treatment process under which the sewage sludge is produced has a considerable bearing on the survival of pathogenic organisms. For example, anaerobic mesophilic digestion (retention 15 days at 35 °C) can reduce 90% of all pathogenic organisms. As with animal wastes, sewage sludges may also contain *E. coli* O157. However, while it is suggested that *E. coli* O157 may remain viable for up to 30 days, there is no evidence of disease transmission when sewage sludge has been used according to current legislation and codes of practice (Carrington *et al.,* 1998; CEN, 1999).

Other non-agricultural organic wastes

In comparison with sewage sludge, much greater quantities of organic wastes from non-agricultural sources are recycled to soils. It is estimated that over 4 million tonnes (dry solids) of a range of industrial by-product organic wastes are applied to agricultural soils in the UK annually. These include wastes from food industries, vegetable pro-cessing, paper industry, textile waste, wood and green plant waste (Carrington *et al.,* 1998). This collection of wastes is extraordinarily variable in quality and form. Table 5.10 gives a range of organic wastes and some of their likely physical, chemical and biological characteristics. The original organic waste may also be altered by composting, whereby bulking agents may be added to increase aeration, as with food wastes or mixed with green wastes, sewage sludges, soybean meal, sugar mill by-products, and many others to improve the C/N ratio (Van Kessel *et al.,* 2000).

Figure 5.6 Manure spreading.

The benefits of land application of non-agricultural organic wastes include increased plant nutrient content and improvements in soil physical and chemical properties, as well as biological and pest and disease control properties (Dissanayake and Hoy, 1999; Zibilske *et al.*, 2000). It is often considered that land application represents an economically and environmentally sustainable alternative to more common methods of disposal of these wastes, such as landfill and incineration. However, potentially detrimental effects of recycling some of these organic wastes to agricultural land may include the presence of pathogenic organisms, potentially toxic elements (metals and organic micropollutants) and salinity (Mullins and Mitchell, 1995).

Detrimental effects of applying non-agricultural wastes to land

Like all organic manures, poorly managed land applications of non-agricultural organic wastes can lead to detrimental effects upon ground and surface water bodies which may have considerable off-site impacts (Chapters 2 and 3) (Box 5.1).

Detrimental in-situ effects upon soil processes by some organic wastes may also occur. Generalisations in this regard are difficult (Table 5.10) as there is considerable physical/chemical and biological variability between wastes. This is further hampered with the use of composts that lack quality assurance from producers as to the presence of metals, pathogens and foreign bodies (such as plastics). Information in this regard is currently missing in many parts of the EU (Moolenaar *et al.*, 1997; Centemero and Corti, 2000).

Table 5.10 General characteristics of a range industrial wastes applied to agricultural land (CEN, 1999).

Waste Source	Abattoirs (Offal)	*Brewing* (Solids)	Papermill sludge	Dairy Milk/whey	Leather Tannery
Dry solids content	Medium	Medium/High	High	Low	Low/High
Nutrient content	High	Medium/High	Low	High	Medium /High
Pathogen content	High	Low	Low	Low	Low/High
pH	6.3	5.8	6.9	5.3	6.7
Cd (mg kg^{-1})	<0.25	<0.25	<0.25	<0.25	<0.25
Cu (mg kg^{-1})	2.4	3.1	32.8	2.4	<1.0
Zn (mg kg^{-1})	9.0	9.9	29.4	1.7	–

Applications of cheese whey (the liquid by-product of cheese production) to agricultural soils has been shown through repeated applications to increase root zone salinity, decrease water infiltration rates (by up to 67%) and when surface applied without incorporation cause a considerable odour problem (Lehrsch and Robbins, 1996). Developing soil salinity and the considerable effects this may have on crop growth is also thought to be a major concern with the repeated application of distillery wastes (Pathak *et al.*, 1999).

One of the most commonly encountered problems associated with the application of organic wastes to agricultural land is the transient imbalance in macronutrients, especially N, in the receiving soil. This may occur with the application of wet olive husks (a by-product resulting during olive oil extraction) (Bedini *et al.*, 1998), corrugated cardboard, wood chips and papermill sludge (PS). An estimated 1 million tonnes PS is produced in the UK per annum, of which just under half is recycled to agricultural land (Aitken *et al.*, 1995). One of the greatest restrictions to the disposal of PS to agricultural soils is the relatively high carbon to nitrogen (C/N) ratio (e.g. 60–225:1), which can result in significant decreases in crop yield, due to nitrogen deficiency (Edwards, 1994; Elvira *et al.,* 1995; Evanylo and Daniels, 1999). The carbon to nitrogen ratio is a characteristic of all organic wastes applied to soils that is often considered to be an integral to the rate of decomposition and release of ammonium and nitrate. When residues having high C/N ratios are added to soils intense competition occurs among microbes for available N, so much so in some cases that higher plants will suffer N deficiency (Figure 5.7) (Brady and Weil, 1999). Furthermore, if there is not sufficient N, the degradation of the added organic substrate will be hindered and potential beneficial breakdown products not released. A C/N ratio in the added organic waste of 20 is thought to be the approximate cut-off value, above which microbial action will not release available N due the decomposition process (Hill and James, 1995).

A common solution to the issue of non-agricultural organic wastes with C/N ratios greater than 20 is to delay planting after amendment or by applying inorganic nitrogenous fertiliser (Logan and Esmaeilzadeh, 1985; Bellamy *et al.,* 1995).

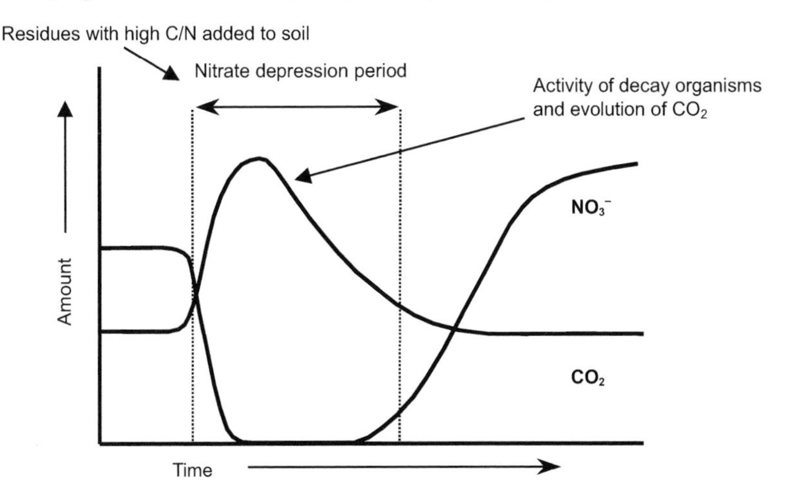

Figure 5.7 The effect of adding organic residues with high C/N ratio on nitrate availability and microbial activity (adapted from Brady and Weil, 1999).

Many of the non-agricultural wastes applied to soils are relatively low in trace metals, certainly when compared to some of the other sources of metals to land (Box 5.2). While it is thought that 20% of all of the Zn recycled to soils through non-agricultural organic wastes is contained in 75,000 t of deinked PS (Carrington *et al.,* 1998), Zn concentrations in PS are often below 40 mg kg^{-1} (Merrington and Madden, 2000). Papermill effluents, produced during the paper production process may also be recycled to land and contain a range of organic micropollutants such as chlorophenols and possibly dioxins. However, while limited research has been undertaken on the potential impacts of these chemicals from this source, it is thought that their environmental impact would be minimal (Kookana and Rogers, 1995; Hawrelak *et al.,* 1999).

In the UK, the utilisation of materials classified as wastes on agricultural land is controlled by the Waste Licensing Regulations (1994), which implement the EU Waste Framework Directive (91/156/EEC) and the Environmental Protection Act (1990). Under the terms of these regulations, the wastes must be recovered to land without endangering human health and without using processes or methods that could harm the environment. In particular, there should not be risk to water, soil, plants or animals and there should be no nuisance through odours. An important feature of these regulations is that the material recovered to land must result in a benefit to agriculture or ecological improvement. In order to assess such benefits, it is necessary to analyse soils where wastes are being applied such that nutrient or physical benefits (e.g. improvement of soil structure) can be demonstrated.

5.4 PRACTICAL SOLUTIONS

Pollution incidents involving organic wastes on the farm are usually caused by poor management, such as:

- spillages;
- run-off due to over-application in the field;
- run-off from yards;
- inadequate storage capacity, structure and management;
- leaking/unknown drainage systems;
- application of slurries when land is frozen or waterlogged.

Various Codes of Good Agricultural Practice (MAFF, 1998a,b,c) have been published by the UK government with the intention of encouraging farmers to manage farm resources effectively, whilst minimising the risk of pollution and keeping within the law. The Code of Good Agricultural Practice for the Protection of Water (MAFF, 1998a) addresses most of the problems associated with organic waste, notably in relation to their threat to watercourses. The Code includes recommendations on the appropriate design and structure of storage and handling facilities. For example, in accordance with the Control of Pollution (Silage, Slurries and Agricultural Fuel Oils) Regulation 1991, silage must be contained within a properly designed and constructed clamp at least 10m away from a watercourse or field drain. The Code also recommends how waste should be handled and disposed of safely; guidelines which have now also been incorporated into a practical farm waste management plan for farmers (ADAS, 1994).

Farm waste handling, storage and disposal

Because of their potential to cause pollution, it is important that all farm wastes are handled and stored safely. The ultimate fate of wastes is usually disposal to land, but prior to this there is often a storage period. Nicholson (1994) identified the objectives of storage as:

- avoidance of water pollution hazards;
- avoidance of damage or contamination to growing crops;
- to optimise use of plant nutrients contained in the waste;
- avoidance of damage to soils by machinery;
- avoidance of requirement for daily spreading.

The way in which the waste is stored and handled depends upon its type. The Code of Good Practice includes the following recommendations for the storage and disposal of different types of farm waste (MAFF, 1998a).

Solid manures

Solid manures (e.g. farmyard manure from traditional covered yards and poultry litter from battery and broiler systems) are less likely to cause pollution than slurries because they are relatively easy to store and handle. However, even with solid wastes there is a liquid fraction, often with high BOD, so they should still be considered a threat. Farmyard manure and other deep litter is typically removed from a farm building using a tractor-mounted fork (Culpin, 1992) and should ideally be stored in a concrete-based structure with up to three walls (2–3 m high), 10 to 15 m wide (MAFF, 1998a) and a below ground tank to store liquid waste. Provided that there is no risk of contaminating adjacent watercourses, an alternative is to put the manure into temporary field heaps.

Several important changes occur during the storage of manures. Aerobic breakdown of organic material leads to a natural composting process, generating carbon dioxide, ammonia and methane. Turning manure heaps during storage will accelerate this process, resulting in less manure to spread, but with elevated nutrient concentrations. In organic farming systems, the storage of manure for at least 3 months prior to application on land is advised (Soil Association, 1999). This storage period, together with turning on two or three occasions will result in manure volume reduction of 50%, leaving in less material to be spread. Nutrients tend to be converted to organic, stable forms, and the manure is more friable, making it easier to spread (Gibbs *et al.,* 2000). Up to 10% of the total N in the manure at the start of the composting/storage phase can be lost by ammonia volatilisation. In addition, manure heaps that are exposed to the elements will lose nutrients in leachate, particularly nitrates and potassium. The speed of incorporation of manures into soil can have a significant effect on the conservation of readily available N in the manures. With slurries often needing to be incorporated within 6 hours of application if 50% of the available N is to be conserved and not lost (Shepherd and Gibbs, 2001). Following storage, the manure can be applied to the land when appropriate by using a spreader such as a rotary flail side-spreader, whilst any accumulated liquid can be spread by tanker or a low-rate irrigation system.

Silage

Silage has traditionally been stored in a clamp or silo, baled silage has become increasingly popular since the late 1970s. The design of a silo must intercept any effluent produced and MAFF recommend that silos have concrete floors which are sealed to prevent liquid escaping and slope from back to front (with a fall of 1 in 75) towards a drain across the front of the silo (Figure 5.8). Effluent should be stored in an appropriate below ground tank and disposed of by tanker.

Patterson and Steen (1994) have shown that an alternative to disposal is to use the effluent as a foodstuff. The study showed that with pigs and cattle silage effluent could make a protein and energy contribution towards production. Big-bale silage is an alternative production method and is claimed to reduce the risk of pollution from effluent by:

- carrying out baling at crop DMs of greater than 30%, thus reducing effluent production;
- containing any effluent produced within the wrapping of individual bales.

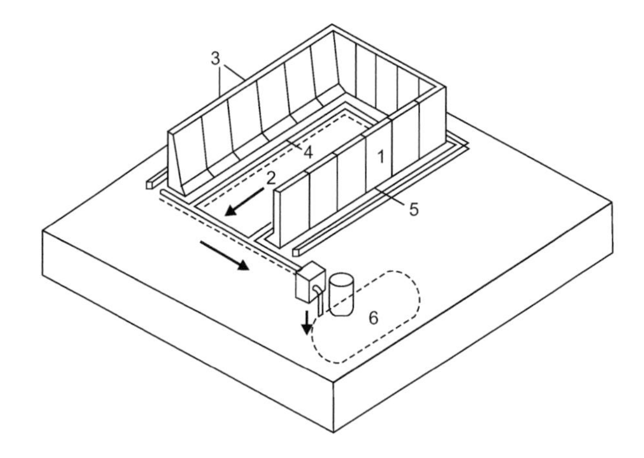

1 Walls built to stand up to loadings in British Standard 5502: Part 22:1987
2 Floor made from suitable concrete mix or hot rolled asphalt
3 Sealant
4 Drains inside the silo walls
5 Channels outside the silo walls
6 Corrosion-resistant tank normally built in one piece with a lockable cover

Figure 5.8 A typical walled silo (MAFF, 1998a).

Slurry

Slurry is usually transferred from housing to store via a reception pit or tank, and is then stored in one of a number of ways: above-ground circular stores, weeping wall stores or earth-banked lagoon stores (MAFF, 1998a).

Above ground circular stores are usually made from coated steel panels, with a depth of up to 6 m. A typical system has a reception pit next to the main store (Figure 5.9) with a pump which moves slurry from the pit to the store. Slurry within the store is usually mixed to prevent gases and odour problems, and to keep the slurry fluid. Slurry is then disposed of from the store by slurry tanker or irrigation system.

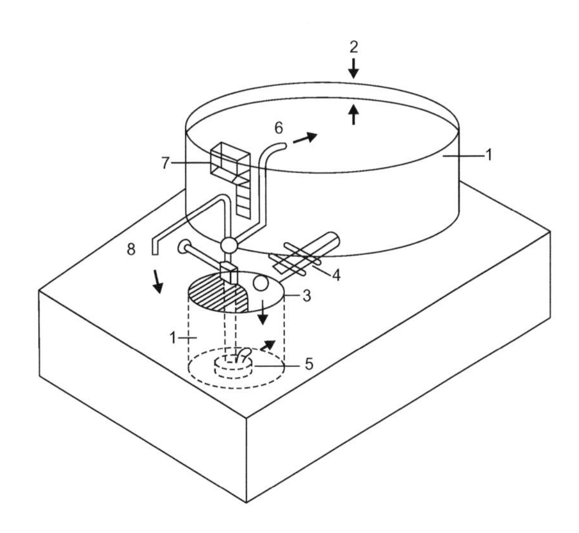

1 Store and reception pit built to British Standard 5502
2 A space of at least 300 mm between the slurry and top of the tank
3 Reception pit large enough for two days slurry production
4 Two values in a line, kept locked when not in use
5 Pump
6 Nozzle to mix slurry
7 Platform for checking and working store
8 Tanker filling point

Figure 5.9 An above-ground circular slurry store (MAFF, 1998a).

Weeping wall stores are above ground structures, typically 2–3 m high built on a concrete base (Figure 5.10). Liquid weeps through gaps in the walls and is collected via channels to a storage tank. As the liquid fraction is allowed to drain away, the solids dry out and have a consistency similar to solid manure. The solid material can be disposed of by spreader, and the liquid by tanker or irrigation.

Earth-banked stores (Figure 5.11) are useful in storing solid, liquid or semi-liquid wastes. They may be built below, above, or partly below ground, and are typically 3–4 m deep. The store can be emptied by tanker or loader and manure spreader depending upon the type of waste. Slurry can also be treated in a number of ways (MAFF, 1998a), including:

- mechanical separation, to give a liquid fraction that can easily be pumped and a solid fraction for spreading as a solid manure;
- anaerobic digestion (treatment without oxygen);
- aerobic digestion (treatment with oxygen).

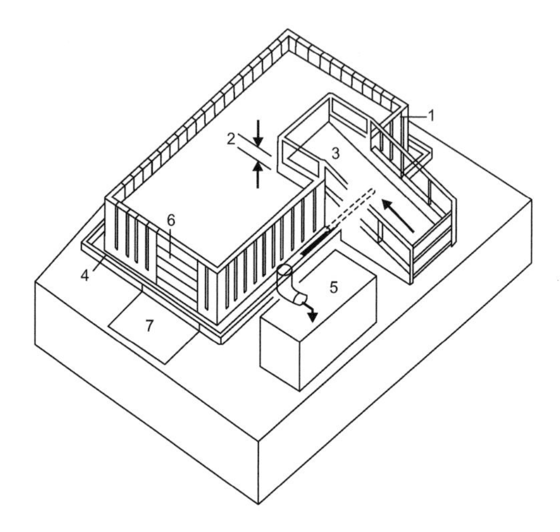

1 Walls built to British Standard 5502 with gaps 25–35 mm wide
2 At least 300 mm space between slurry and top of panels
3 Filling ramp with safety rails
4 Base extended to give channel round store
5 Tank built to BS 5502 of a size to suit farm circumstances
6 Panels that can be taken out to empty the store
7 Hard area for machinery to get to the store

Figure 5.10 A weeping wall slurry store (MAFF, 1998a).

Dirty water

Dirty water is produced on the farm by cleaning work, including yard and parlour washing, and the run-off from open areas (e.g. stock yards and outdoor silos). It can therefore be contaminated with a range of materials, including livestock faeces and urine, silage effluent, milk and parlour cleaning chemicals, with a broad range of BODs and other polluting effects.

Dirty water is usually stored in either an earth-banked lagoon or below ground store. A simple way of separating any solids from this liquid waste is to have a series of settlement tanks or barrier ditches separated by H-section overflow pipes. Disposal of liquids is usually by tanker or low-rate irrigation systems. Low-rate irrigation systems pump the dirty water from the settlement tank or lagoon directly to the land via an electric pump, small bore pipes and sprinkler (or travelling irrigator) in the field.

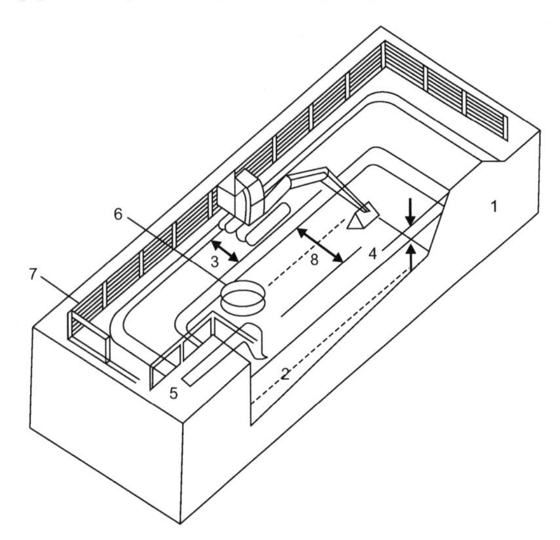

1 Embankments to suit the site
2 Floor slopes slightly towards the filling point
3 If it is emptied from the top, banks are built to take the weight of machine
4 At least 750 mm space between surface of the slurry and the top
5 Filling point
6 Strainer to remove excess liquid to be removed by tanker
7 Safety fence
8 Width to suit the reach of the machine

Figure 5.11 An earth-banked store suitable for solid, liquid or semi-liquid wastes (MAFF, 1998a).

Good agricultural practice

The ADAS Farm Waste Management Plan

The ADAS Farm Waste Management Plan (ADAS, 1994 comprises five steps for farmers to take (Box 5.4)):

STEP 1 Identifies how much land is available for spreading manures and where manure should not be spread. A map is drawn of the farm which shows fields, watercourses and boreholes, springs or wells. Areas are identified where manures and slurries should never be spread. These include:

- areas within 10 m either side of ditches and watercourses;
- areas within 50 m of any spring, well, borehole or reservoir;

- very steep slopes where run-off is a high risk throughout the year;
- any areas where spreading is not permissible, such as a Site of Special Scientific Interest (SSSI) or ESA.

STEP 2 *Identifies restrictions on when manure should be spread*
This step identifies those areas of the farm which may not be suitable for spreading manure at certain times of year e.g. due to waterlogging etc. The remaining areas are suitable for spreading manure at any time as long as the ground is not frozen or the ground is so wet as to cause damage to the soil. A colour key is used to identify the areas within the farm with their risk associated (Figure 5.12).

STEP 3 *Calculates minimum area of land needed for spreading*
From the livestock numbers and the number of months cattle are housed, the minimum area of land needed for spreading animal manures and slurries is calculated (Table 5.12). This ensures that the MAFF Code of Good Agricultural Practice guideline of no more than 250 kg N ha^{-1} $year^{-1}$ is not exceeded (Chapter 2).

STEP 4 *Gives guidance on applying sewage sludge and other organic wastes*
This step accounts for land required if sewage sludge or other organic wastes are spread on the farm.

STEP 5 *Calculates storage requirements for slurry and dirty water*
If the farm produces slurry and dirty water, a calculation is carried out to ensure that the existing storage capacity is needed to minimise the risk of pollution. This calculation is based on existing storage capacity, months of the year when livestock are housed, and an assessment of the volumes of waste produced.

Table 5.11 ADAS Farm Waste Management Plan areas of high risk for slurry or manure spreading (ADAS, 1994).

Risk	Conditions leading to risk
Very High	Fields or parts of fields next to a watercourse, spring or borehole when the surface is severely compacted or waterlogged.
Very High	Fields or parts of fields that are likely to flood sometime in most winters.
Very High	Fields or parts of fields next to a watercourse, spring or borehole when the soil is at field capacity and there is a steep slope or a moderate slope and a slowly permeable soil.
High	Fields or parts of fields next to a watercourse, spring or borehole when the soil is at field capacity and there is a moderate slope and a well drained soil or a slight slope and a slowly permeable soil.
High	All fields or parts of fields with effective pipe or mole drains. Shallow soils over gravel or rock, e.g. limestone, chalk, slates and shales.

Table 5.12 Extract of table from *ADAS Farm Waste Management Plan* to calculate the areas needed to spread slurry during months when all livestock are housed (ADAS, 1994; MAFF/WOAD, 1998).

Stock unit on slurry based system	Number of stock	Proportion of waste as slurry	Hectares needed per unit per month	Area needed
1 Dairy Cow	x	x	0.031	=
1 Suckler Cow		x	x 0.024	=
1 Beef Bullock		x	x 0.011	=
1 Follower	x	x	0.009	=
		TOTAL AREA NEEDED PER MONTH		=

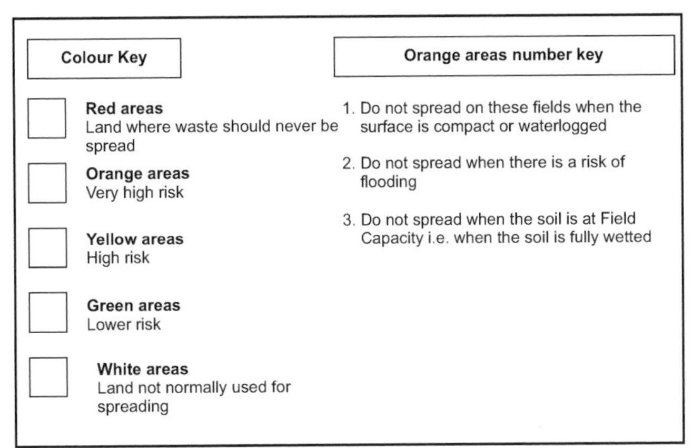

Figure 5.12 Cut out colour key to aid in the identification of areas of high risk in the *ADAS Farm Waste Management Plan* (ADAS, 1994, MAFF/WOAD, 1998).

Box 5.4 Farm waste management plan case study

Background

Laurel Farm is a medium-sized holding of 144 ha located in Dorset (average rainfall of 450 mm). It is a traditional mixed dairy farm with approximately 40 ha of arable land producing forage maize and spring/winter barley in rotation with 60 ha of temporary grassland. The remaining land is permanent pasture.

The herd comprises 120 Friesians with 30 followers. During the winter (approximately 6 months) all livestock are housed in deep litter barns with access to outdoor yards for feeding (2000 m^2). Slurry from the yards is scraped daily by tractor into a weeping-wall store from which dirty water (plus run-off from the yard and dairy washings) drains into a below-ground tank connected to a low rate irrigation system.

When the stock are housed during the winter, the slurry store often needs emptying every 2–3 weeks. The slurry is scooped by tractor-mounted bucket into a manure spreader and, depending upon weather conditions, spread onto suitable land. The deep litter barns are cleaned out twice – once in mid-winter and then again when the livestock return to pasture. The manure may be heaped and stored, or spread immediately depending upon prevailing conditions.

It is planned to increase the herd size during 1996/97 to 200 cows with 50 followers. While the available buildings and yard are adequate for a larger herd size, the existing waste management system is old and may not be adequate to cope with the increased herd size. The farmer has some capital funds available, but wishes to minimise costs by retaining as much of the existing system as possible. In order to assess the viability of the proposal a consultant was asked to produce an appropriate report, including a farm waste management plan.

The assessment

The first stage in assessing the proposal was to conduct a thorough audit of existing facilities and farming practice.

Using the ADAS Farm Waste Management Plan (ADAS, 1994), a map was produced to identify those areas on the farm suitable for spreading organic waste. In summary, this revealed:

Field Descriptions	Note	Area (ha)
Permanent pasture designated as a SSSI and subject to management agreement	Application is prohibited	7.8
Fields adjacent to a river and susceptible to winter flooding	Very high risk of pollution	24.3
Gently sloping land with recently installed pipe-drains discharging into a small ditch	High risk of pollution	15.0
No identifiable factors increasing the risk of pollution	Lower risk of pollution	96.9

The only fields where slurry and manure are not being spread at present are the 7.8 ha subject to SSSI management agreement. The total area needed in winter for spreading the greater volumes of manure and slurry produced on the farm when the herd size is increased was calculated as follows:

Stock Unit	**Number of Stock units**	**Months housed**	**Hectares needed per stock unit**	**Total area needed (ha)**
1 dairy cow	200	x 6	0.031	= 37.2
1 follower	50	x 6	0.009	= 2.7
			Minimum land needed	**= 39.9**

It was concluded that there is sufficient land available to dispose safely of the extra waste produced, subject to the following GENERAL conditions:

- continue to avoid applying wastes to the SSSI;
- do not apply wastes in winter time to those fields identified as susceptible to flooding or those that have active field drains;
- never apply wastes within at least 10 m of either side of any watercourse on the farm, including ditches;
- never apply wastes when the ground is frozen, or the fields are so wet that tractor-drawn machinery will damage the soil.

There are currently three types of organic waste produced on the farm: slurry from yard scrapings, dirty water, and farm yard manure from the deep litter housing.

These wastes differ in their pollution potential. The solid manure produced by the deep litter system is potentially the least polluting and at Laurel Farm is often spread in the late winter/early spring prior to the sowing of spring cereals. If the manure must be stored at all it is stacked in temporary field heaps well away from field drains and other watercourses.

This was considered to be an acceptable practice that can be continued subject to the following specific conditions:

- Do not exceed an application of 40 $t ha^{-1}$ of fresh farm yard manure (equivalent to 250 $kg N ha^{-1}$) on any area within a 12 month period. This includes land used for growing maize.
- In order to maximise their fertiliser value and reduce the risk of nitrate leaching, manures should be applied as a complement to inorganic fertiliser applications and according to crop need.

Special attention was given to the management of slurry and dirty water (including silage effluent) on the farm since these are potentially the most polluting wastes.

Dirty water storage had been identified as a problem by the farmer because it was reported that the underground store had been over-flowing during heavy rainfall. Inspection of the farm buildings, however, revealed that drainpipes from the farm building roofs were broken, resulting in virtually all clean water run-off from the roofing (some 1200 m^2) entering the below-ground store.

It was recommended that this clean rain water is separated from dirty water by repairing the drainage systems of the farm buildings. It is anticipated that once this work is carried out, the existing below-ground store should have enough capacity for the amount of dirty water produced.

Using the equations from the ADAS Farm Waste Management Plan (ADAS, 1994), the minimum area needed for spreading the increased volume of dirty water in winter was also calculated as follows:

Volume of total run-off (m^3) = Area (m^2) x Typical winter rainfall (mm)/1000

Run-off from open yards:	2000 x (450 x 0.6)/1000	= 540 m^3
Run-off from uncovered silage clamp:	400 x (450 x 0.6)/1000	= 108 m^3
Run-off from weeping-wall slurry store:	200 x (450 x 0.6)/1000	= 54 m^3

Volume of parlour washings (m^3) = Number of cows x 3.24

200 x 3.24 = 648 m^3

Total volume of dirty water produced = 1350 m^3

Minimum area needed for spreading dirty water in winter = (Total volume of dirty water/1000) x 4

(1350/1000) x 4 = **5.4 ha**

Again it was concluded that there is sufficient land available to dispose safely of the dirty water produced during the winter months, subject to the following specific conditions:

- apply no more than 5 mm (5 l m^{-2}) per hour of dirty water through low volume sprinklers;
- move sprinklers regularly to suit conditions.

According to the MAFF Code of Good Practice for the Protection of Water (MAFF, 1998a), weeping-wall stores are considered suitable for cattle slurry with a lot of straw bedding in it and thus appropriate to retain at Laurel Farm.

However, it was concluded that the current store would not be large enough for the increased herd size and that it should be enlarged and improved to ensure 4 months storage capacity at BS5502 standards. The application of slurry should also be subject to the following specific condition:

- do not exceed an annual application of 50 m^3 ha^{-1} of undiluted cattle slurry (equivalent to 250 kg N ha^{-1}) on any area within a 12 month period. This includes land used for growing maize.

With the improvements outlined, plus strict adherence by all farm staff to the prepared farm waste management plan and regular inspection of the storage facilities (i.e. to ensure that they are not deteriorating and that any repairs are carried out immediately), the consultant considered that the herd size could be safely expanded at Laurel Farm.

5.5 NON-AGRICULTURAL ORGANIC WASTES

Whilst it is estimated that some 60% of agricultural land in England and Wales is suitable for accepting sewage sludge, the actual area available for agricultural disposal of sewage sludge is constrained by factors such as:

- public acceptability, and perceived risks;
- risk of surface water pollution;
- cropping patterns;
- the willingness of farmers to accept sludge (RCEP, 1996; Renner, 2000);
- the adjacency of land to sewage works, only 10% of farmland is within acceptable transport distance of sewage treatment works.

Taking account of all these constraints only 0.5% of agricultural land in England and Wales is currently used for sludge disposal (RCEP, 1996). Although the application of sewage sludge to agricultural land may be potentially beneficial, care must be taken to avoid the risk of potentially toxic elements, such as trace metals, increasing to unacceptable levels in the soil. However, this is only likely to occur where repeated applications are made to relatively small areas of land and the potential impacts of such disposal practice in the long-term are not yet fully understood (Chaney *et al.*, 2000; Martinez *et al.*, 2001). It is more likely that, given the current levels of potentially toxic elements in sewage sludges in the UK, the greatest factor limiting application of sewage sludge will often be the N limit of 250 kg ha^{-1} $year^{-1}$ (MAFF, 1998a). Nevertheless, the application of sewage sludge is subject to a range of controls and guidelines. In the UK, these measures have effectively led to; legal restrictions on the amount of Zn, Cu, Ni, Cd, Pb, Hg and Cr applied in sludge and accumulating in soil, and recommended limits (not legally set) on Mo, Se, As and F in sludge and soils.

Within all EU Member States, concentrations of heavy metal contaminants in sewage sludge applied to agricultural land are strictly controlled under EC Directive 86/278 on the Protection of the Environment and in Particular of the Soil, when Sewage Sludge is used in Agriculture (EC, 1986). The EC Directive adopts a two-pronged approach by specifying ranges of concentrations of metals which Member States are required to use as the basis for:

- setting limits for metal levels in soils (of pH 6–7) which must not be exceeded as the result of using sewage sludge; and
- controlling the amounts of metals added to land in sewage sludge.

The EC Directive is implemented in the UK by the Sludge (Use in Agriculture) Regulations 1989 (HMSO, 1989) (this document was to be revised in 2000, but is yet to be published) which makes it an offence to knowingly allow sludge to be applied to farmland when it does not meet legal requirements. Further guidance is available in the Code of practice for agricultural use of sewage sludge (DoE, 1996) which amongst other things identifies the maximum permissible concentrations (MPCs) in soil of different metals in relation to soil pH (Table 5.13). Importantly, these values are currently being revised, but have yet to be published (EU, 2000; Renner, 2000).

Avoiding pollution problems by potentially toxic elements in sewage sludge

A comparison of the MPCs of metals in soil at pH 6–7 set by the UK with some other EU Member States and the United States is shown in Table 5.13. Although UK MPCs are some of the least stringent in the EU (a number of Member States chose to adopt lower limits than required by the EC Directive), it is important to note that UK regulations do take account of the fact that the availability of metals for crop uptake varies with soil pH (Alloway, 1995). Since most metals (with exceptions such as selenium and molybdenum) become more available for uptake in acid soils different MPCs are set for different pH values (Table 5.13) and sewage sludge application is prohibited on soils with a pH below 5.0 (DoE, 1996). The maintenance of soil pH in fields regularly receiving sludge is therefore very important, especially where trace metal levels are already at, or close to, the MPC for a particular pH range.

A comparison with the values set in the United States reveals a differing philosophy to that of the EU member states in regard to the setting of metals limit values (US EPA, 1993; Renner, 2000). The European approach is precautionary in the light of what is thought to be a lack of clear scientific evidence as to what levels potentially toxic elements from sludges applied to soils may become toxic. Indeed, it is expected that the revised EU sludge directive will continue this precautionary approach and attempt to balance the addition of metals in sewage sludges with off-takes, such as crop removal, and therefore prevent the accumulation and build up of metals within the plough layer. In the United States, scientist believe that there is enough evidence available to indicate there would be negligible detrimental threat from metals in sewage sludge if applied within the suggested metal limits (Renner, 2000). Importantly, there are detractors to each of these approaches and only time will tell who has been over cautious or far too permissive (McBride, 1995; Schmidt, 1997; Chaney *et al.*, 1999).

It must also be remembered that many other materials contain one or more of the contaminants present in sewage sludge and so are also subject to the MPCs in Table 5.13. For example, the main contaminants in animal manures are Cu and Zn that derive from feed additives, and it is advisable for farmers to monitor their levels in soils receiving repeated applications of pig or poultry manure. The mixing of sludge with other wastes or products should still conform to the MPCs above, both before and after, i.e. the mixing of sludges with other wastes for the purpose of diluting pollutants is prohibited (EU, 2000).

The UK MPCs for potentially toxic elements listed in Table 5.13 were set on the basis of the potential direct toxic effects on crops with due consideration to aspects of human and animal health. Although the potentially toxic effects of metals on soil organisms and associated microbiological processes have previously caused the revision and lowering of the guide value for Zn, from 300 to 200 mg kg^{-1} (MAFF, 1993). It is stated that the use of sludge is to be carried out to in a way that minimises detrimental effects on; human plant and animal health, the quality of surface and groundwaters, the long-term quality of soil including the soil microbial population (EU, 2000).

Table 5.13 A comparison of maximum permissible total selected metal concentrations (mg kg^{-1}) in sludge amended soil set in EC Directive 86/278/EEC and those administered by individual EC Member states and also the United States (MAFF, 1993; Renner, 2000).

	Cd	Cu	Pb	Zn
European Community	1–3	50–140	50–300	15–300
Spain	1	50	50	150
UK*	3	135	300	200
Norway	1	50	50	150
Sweden	0.5	40	40	100
United States**	20	750	150	1400

* Values are for soil pH 6–7, other values apply at pH 5–6 and >7.
** Calculated from maximum cumulative pollutant loading limits mixed into the soil plough layer.

Avoiding crop contamination by pathogens

Good management practices can also help to avoid the risk of crop contamination by pathogens, such as *E. coli*, *Salmonella* and *Cryptospiridium*, commonly found in farm wastes and sewage sludge. Pre-treatment (fermenting, aerating or composting) of farm wastes prior to application can greatly reduce the risk of pathogen contamination (Jones, 1999). Schechtner (1992) conducted trials on the effects of aeration on the physical and chemical properties of slurries and demonstrated that the heat generated by aeration was effective in killing some pathogens.

The Code of Good Practice for the Agricultural Use of Sewage Sludge (DoE, 1996) and CEN (1999) also provide guidance and recommendations on:

- Effective sewage treatment processes, such as high temperature treatment and anaerobic digestion, which can significantly reduce the risk of health hazards when the resultant sludge is applied to agricultural land.
- Sludge spread on the soil surface not only dries more quickly, thus shortening the survival times of pathogens, but is also exposed to oxygen and ultra-violet radiation in sunlight. The surface spreading of treated sludge on grazed grassland was banned at the end of 1998, if applied it must now be deep injected.
- Acceptable uses (subject to conditions) for sewage sludge. Organic wastes should not, for example, be applied directly to crops, especially if they are to be eaten raw. This is intended to reduce the immediate risk of micro-organisms being transmitted directly to animals or humans, but may allow some pathogens to survive for longer in the soil (Carrington *et al.*, 1998). Treated sludge can only be applied to land used to grow vegetables at least 12 months prior harvest of the crop. For salad vegetables this increases to 30 months (Hickman *et al.*, 1999).

It is expected that the revised Sludge (Use in Agriculture) Regulations will bring in the 'Safe Sludge Matrix' concept. The Matrix is an agreement between major stakeholders, such as ADAS, the water companies and the British Retail Consortium and introduces a number of major changes in the way in which sludge is used on agricultural land. It was brought about to ensure a minimum standard of sustainable sewage sludge recycling to land and comprises of a table of crops and levels of sewage sludge treatment, with guidance for suitable levels of sludge treatment for applications soils growing certain crops (Table 5.14).

A number of other countries have adopted a more cautious approach to the risk of pathogen contamination from sewage sludge spread on land (RCEP, 1996). In Germany and the USA only treated sewage can be applied, which will also be the case from 2002 onwards in the UK (SEPA, 1998). While in New South Wales (Australia) and the USA the public are excluded for over 6 months from land to which sludge has been applied.

Table 5.14 The 'safe sludge matrix' (Hickman *et al.*, 1999).

Cropping Category	Untreated sludges	Treated sludges	Enhanced treated sludges*
Fruit	✗	✗	✓
Salad	✗	✗	✓
Vegetables	✗	✗	✓
Horticulture	✗	✗	✓
Combinable and animal feed crops	✗	✓	✓
Grass, grazing	✗	✗	✓
Grass, silage	✗	✓	✓
Maize, silage	✗	✓	✓

✗ = Application not allowed.
✓ = Applications must comply with Code of Practice and crop and harvest restrictions.
* Including heat treated and other methods.

5.6 ALTERNATIVE TECHNOLOGIES FOR FARM WASTE TREATMENT

In an examination of the problems of pollution caused by livestock production systems, two further practical solutions to the potential problems of dealing with organic farm wastes may be identified – biogas production and reed bed treatment (RBT).

Biogas production

An alternative to applying slurry and manure to land is to anaerobically digest the organic materials with micro-organisms to produce biogas; a mixture of methane (55–65%) and carbon dioxide (35–45%). Wase and Thayanithy (1994) concluded that with advances in technology available '...the currently advocated process is stable, well researched, and economically viable when operated under suitable conditions'.

However, there are currently relatively few (≈25) anaerobic digestors operating on UK farms (RCEP, 1996). The main obstacle to greater uptake remains the economics of digestion. One solution might be the installation of centralised digestors serving a number of local farms. These already operate in Denmark and are currently being investigated in the UK. A preliminary study has shown that an anaerobic digestor receiving slurry from farmers within a range of about 8 miles and linked to a combined heat and power station would cost £4–5 million and could yield 1MW of electricity for sale to the national grid or local consumers; similar quantities of hot water for local householders or businesses; a liquid manure for return to local farmers; and a compost material for sale (RCEP, 1996). Although a plant of this nature is unlikely, at least for the foreseeable future, to be profitable without public subsidy, it would bring significant environmental benefits, especially in those areas where there is increasing risk of serious environmental pollution from farm wastes.

Reed bed treatment

An alternative way of treating dirty water, such as dairy washings and yard run-off, is to use a RBT system. This is an artificially constructed wetland usually planted with Common Reed (*Phragmites australis*) through which the dirty water slowly trickles. The reeds not only absorb nutrients such as nitrogen and phosphorus, but also have the ability to transfer oxygen down through their stems and out via their root system into the surrounding rhizosphere. This increases the capacity of the system for the aerobic bacterial decomposition of organic pollutants (e.g. milk, urine and faeces), as well as encouraging the proliferation of a wide range of aquatic organisms, some of which directly utilise additional pollutants. Biddlestone *et al.* (1994) discuss a composting and reed bed system on a dairy farm with 140 cattle which treats parlour washings and yard waste water, and results in substantial reductions in BOD and suspended solids.

There are numerous designs of RBT system for treating sewage, industrial effluents and highways run-off, as well as agricultural wastes (WRc, 1996). Their main benefits, compared to the conventional treatment of dirty water in tanks and lagoons, are claimed to be low capital cost, very effective water treatment, minimal (if not enhanced) visual impact, and little smell.

It is clear from the above discussion that agricultural and non-agricultural wastes may be valuable sources of plant nutrients and soil conditioner/ improver. Yet, potential problems associated with their handling, storage and disposal may have broad implications for the environment beyond the farm. Effective organic waste management and good agricultural practice may entail:

- knowing and valuing the organic waste nutrient content;
- using this information to balance nutrient inputs and removals, such as crop offtake;
- reduce losses of organic wastes from storage and anima housing;
- apply the wastes evenly and incorporate into soil rapidly;
- timely applications of organic wastes have a significant influence on nutrient loss, i.e. avoid late summer or early autumn applications (Shepherd and Gibbs, 2001).

REFERENCES

ADAS (1994) *Farm Waste Management Plan: the ADAS Step-By-Step Guide for Farmers*. Ministry of Agriculture, Fisheries and Food, London.

Aitken, M.N., Lewis, J.G. and Evans, B. (1995) Effects on soil fertility from applying paper mill sludge to agricultural land. *Soil Use and Management* **11**, 152–153.

Alloway, B.J. (1995) *Heavy Metals in Soils*. 2nd Edition. Blackie Academic and Professional, London.

Alloway, B.J. and Ayres, D.C. (1993) *Chemical Principles of Environmental Pollution*. Blackie Academic and Professional, Glasgow.

Arnold, S.D. and Meister, E.A. (1999) Dairy feedlot contribution to groundwater contamination: a preliminary study in New Mexico. *Journal of Environmental Health* **62**, 16–19.

Barry, G.A., Gardner, E.A. and Bloesch, P. (1998) Studies on reuse of biosolids on Queensland agricultural lands. In: *Watertech, Proceedings Of The Australia Water And Wastewater Association Conference*, Brisbane, Qld. (CD ROM).

Bellamy, K.L., Chong, C. and Cline, R.A. (1995) Paper sludge utilization in agriculture and container nursery culture. *Journal of Environmental Quality* **24**,1074–1082.

Bedini, S., Privitera, I., Filippi, C. and Nuti, M.P. (1998) Microbiota in soil treated with non-fermented wet olive husks. *Rivista-di-Agronomia* **4**, 264–270.

Biddlestone, A.J., Gray, K.R. and Thayanithy, K. (1994) Composting and reed beds for aerobic treatment of livestock wastes. In: *Pollution in Livestock Production Systems* (Eds I. Ap Dewi, R.F.E., Axford I., Marai, M. Fayez, and H. Omed). CAB International, Wallingford, pp. 345–360.

Brady, N.C. and Weil, R.R. (1999) *The Nature and Properties of Soil*. 12th Edition. Prentice Hall, New Jersey.

Brookes, P.C. and McGrath, S.P. (1984) The effects of metal toxicity on the size of soil microbial biomass. *Journal of Soil Science* **35**, 341–346.

Carrington, E.G., Davies, R.D., Hall, J.E., Pike, E.B., Smith, S.R. and Unwin, R.J. (1998) Review of the Scientific Evidence Relating to the Controls on the Agricultural Use of Sewage Sludge. Part 2 – Evidence Since 1989 Relevant to Controls on the Agricultural Use of Sewage Sludge. WRc report No. DETR 4454/4. WRc Medmenham.

CEN (1999) Recommendations To Preserve And Extend Sludge Utilisation And Disposal Routes. Comite Europeen de Co-ordination des Normes (European Committee for Co-ordination of Standards). CEN TC 308.

Centemero, M. and Corti, C. (2000) Technical characteristics of compost for sustainable agriculture. *Informatore-Agrario* **6**, 33–39.

Chaney, R.L., Ryan., J.A. and Brown, S.L. (1999) Environmentally acceptable endpoints for soil metals. In: *Environmental Availability of Chlorinated Organics, Explosives, and Metals in Soils* (Eds W.C. Anderson, R.C., Loehr and D. Reible). *Am. Acad. Environ. Eng*., Annapolis, MD, pp. 111–155.

Chaney, R.L., Brown, S.L., Angle, J.S., Stuczynski, T.I., Daniels, W.L., Henry, C.L., Siebielec, G., Li, Y.M., Malik, M., Ryan, J.A. and Compton, H. (2000) In situ remediation/reclamation/restoration of metals contaminated soils using tailor-made biosolids mixtures. In: *Symposium on Mining, Forest and Land Restoration: The Successful Use of Residuals/Biosolids/Organic Matter for Reclamation Activities*. Rocky Mountain Water Environment Association, Denver, Co, July 17–20th.

Conway, G.R. and Pretty, J.N. (1991) *Unwelcome Harvest: Agriculture and Pollution*. Earthscan, London.

Culpin, C. (1992) *Farm Machinery*. Blackwell Scientific Publications, Oxford.

Dampney, P. (1995) Good timing helps bank balances. *Farmers Weekly (Forage and Grassland Supplement)*, 17th February 1995, 20–25.

DoE (1993) *River Quality Draft Regulation: The Surface Waters (Fisheries Ecosystem) (Classification) Regulations*. Department of the Environment, London.

DoE/DH (1990) *Crytosporidium in Water Supplies*. Report of the Group of Experts, Department of the Environment and Department of Health. HMSO, London.

DoE (1996) *Code Of Practice For Agricultural Use Of Sewage Sludge*. HMSO, London.

DETR (2001) Chapter 3 Inland Water Quality and Use. In: *Digest of Environmental Statistics*. Department of the Environment, Transport and the Regions, London.

Dissanayake, N. and Hoy, J.W. (1999) Organic material soil amendment effects on root rot and sugarcane growth and characterization of the materials. *Plant-Disease* **11**, 1039–1046.

EC (1986) EC Directive 86/278/EEC on the Protection of the Environment and in Particular of the Soil, when Sewage Sludge is used in Agriculture. Official Journal of the European Communities No. L181/8 (4/7/86).

Edwards, J.H. (1994) Direct land application of waste paper. *Biocycle,* April, 69–72.

Edwards, A.C. and Withers, P.J.A. (1998) Soil phosphorus management and water quality: A UK perspective. *Soil Use and Management* **14**, 124–130.

Elvira, C., Mato, S. and Nogales, R. (1995) Changes in metal extractability and organic matter fractions after vermicomposting of sludges from a paper mill industry and wastewater treatment plant. *Fresenius Environmental Bulletin* **4**, 503–507.

Environment Agency (2001a) *Water Quality*. URL: http://www.environment-agency.gov.uk/subjects/waterquality/134499/?version=1

Environment Agency (2001b) *Sewage Sludge*. URL: http://www.environment-agency.gov.uk.sewage sludge

EU (2000) Working document of sludge, 3rd Draft. European Union, Brussels. EnV.E.3/LM.

Evanylo, G.K. and Daniels, W.L. (1999) Paper mill sludge composting and compost utilization. *Compost-Science and Utilization* **7**, 30–39.

Foy, R.H. and Kirk, M. (1995) Agriculture and water quality: A Regional Study. *Journal of the Chartered Institute of Water and Environmental Management* **9**, 247–256.

Foy, R.H., Smith, R.V., Smyth, D., Lennox, S.D. and Unsworth, E.F. (1994) The impact of climatic and agricultural variables on the frequency of silage pollution incidents. *Journal of Environmental Management* **41**, 105–121.

Gibbs, P.A., Parkinson, R.J., Fuller, M.P. and Misselbrook, T. (2000) Enhancing the effective utilisation of animal manures on-farm through compost technology. In: *Agriculture and Wastes. Management for a Sustainable Future* (Eds T. Petchey and B. D'Arcy). SAC, Edinburgh, pp. 63–72.

Gimeno-García, E., Andreu, V. and Boluda, R. (1996) Heavy metals incidence in the application of inorganic fertilizers and pesticides to rice farming soils. *Environmental Pollution* **92**, 19–25.

Giller, K.E., Witter, E. and McGrath, S.P. (1998) Toxicity of heavy metals to microorganisms and microbial process in agricultural soils: a review. *Soil Biology and Biochemistry* **30**, 1389–1414.

Glendinning, J.S. (1999) *Australian Soil Fertility Manual*. CSIRO Publishing, Collingwood, Australia.

Haigh, P.M. (1997) Silage dry matter content and predicted effluent production in England and Wales 1984–1994. *Journal of Agricultural Engineering Research* **66**, 63–77.

Hall, J.E. (1992) Treatment and use of sewage sludge. In: *The Treatment And Handling Of Wastes* (Eds A. Bradshaw, D. Southwood and R. Warner). Chapman and Hall, London, pp. 63–82.

Hawrelak, M., Bennett, E. and Metcalfe, C. (1999) The environmental fate of the primary degradation products of alkylphenol ethoxylate surfactants in recycled paper sludge. *Chemosphere* **39**, 745–752.

Hickey, C.W., Golding, L.A., Martin, M.L. and Crocker, G.F. (1999) Chronic toxicity of ammonia to New Zealand freshwater invertebrates: a mesocosm study. *Archives of Environmental Contamination and Toxicology* **37**, 338–351.

Hickman, G.A.W., Chambers, B.J. and Pierrepont, P.J. (1999) Addressing customer concerns – development of the ADAS matrix for sustainable application of sewage sludge to agricultural land. In: *Agriculture and Wastes Management for a Sustainable Future* (Eds A.M. Petchey, B.J. D'Arcy and C.A. Frost). SAC Edinburgh, pp. 4–10.

Hill, R.L. and James, B.R. (1995) The influence of waste amendments on soil properties. In: *Soil Amendments and Environmental Quality* (Ed. J.E. Rechcigl). CRC Pres, Inc, Boca Raton, Fla, pp. 311–325.

HMSO (1989) Sludge (Use In Agriculture) Regulations 1989, S.I. 1989 No. 1263 As Amended by the Sludge (Use in Agriculture) Regulations 1990, S.I. 1990 No. 880. Her Majesty's Stationery Office, London.

HMSO (1991) Control of pollution (Silage, Slurry and Agricultural Fuel Oil) Regulations 1991, S.I. 1991 No. 324. Her Majesty's Stationery Office, London.

HMSO (1993) Agriculture in the United Kingdom 1993. MAFF, SOAFD, DANI, WO, Her Majesty's Stationery Office, London.

Huysman, F., Verstraete, W. and Brookes, P.C. (1994) Effect of manuring practices and increased copper concentrations on soil microbial populations. *Soil Biology and Biochemistry* **26**, 103–110.

Jackson, D.R. and Smith, K.A. (1997) Animal manure slurries as a source of nitrogen for cereals: effect of application time on efficiency. *Soil Use and Management* **13**, 75–81.

Johnes, P.J., Moss, B. and Phillips, G. (1996) The determination of total nitrogen and total phosphorus concentrations in freshwaters from land use, stock headage and population data: testing a model for use in conservation and water quality management. *Freshwater Biology* **36**, 451–473.

Jones, D.L. (1999) *Escherichia coli* O157 in the environment. *Soil Use and Management* **15**, 76–83.

Kookana, R.S. and Rogers, S.L. (1995) Effects of paper mill effluent disposal on soil. *Reviews of Environmental Contamination and Toxicology* **42**, 13–64.

Lehrsch, G.A. and Robbins, C.W. (1996) Cheese whey effects on surface soil hydraulic properties. *Soil Use and Management* **12**, 205–208.

Lennox, S.D., Foy, R.H., Smith, R.V., Unsworth, E.F. and Smyth, D.R. (1998) A comparison of agricultural water pollution incidents in Northern Ireland with those in England and Wales. *Water Research* **32**, 649–656.

LeRiche, H.H. (1968) Metal contamination of soil in the Woburn Market Garden experiment resulting from the application of sewage sludge. *Journal of Agricultural Science (Camb.)* **71**, 205–208.

Logan, T.J. and Esmaeilzadeh, H. (1985) PS evaluated for use on cropland. *Ohio Report*, March–April, 22–25.

MAFF (1993) *Review of the Rules for Sewage Sludge Application to Agricultural Land: Soil Fertility Aspects of Potentially Toxic Elements*. Ministry of Agriculture, Fisheries and Food, London.

MAFF (1998a) *Code of Good Agricultural Practice for the Protection of Water*. Revised 1998. Ministry of Agriculture, Fisheries and Food, London.

MAFF (1998b) *Code of Good Agricultural Practice for the Protection of Air*. Revised 1998. Ministry of Agriculture, Fisheries and Foods, London.

MAFF (1998c) *Code of Good Agricultural Practice for the Protection of Soil*. Revised 1998. Ministry of Agriculture, Fisheries and Food, London.

MAFF/WOAD (1998) *Farm Waste Management Plan. A Step by Step Guide to Farmers*. Ministry of Agriculture, Fisheries and Food, London.

MAFF (2000) *Towards Sustainable Agriculture: Pilot Set of Indicators*. Ministry of Agriculture, Fisheries and Food, London, p. 30.

Martinez, C.E., Jacobson, A. and McBride, M.B. (2001) Thermally indiced changes in metal solubility of contaminated soils is linked to mineral recrystallization and organic matter transformations. *Environmental Science and Technology* **35**, 908–916.

Mason, C.F. (1996) Water pollution biology. In: *Pollution, Causes, Effects And Control*. (Ed. R.M. Harrison). 3rd Edition. The Royal Society of Chemistry, Cambridge, pp. 66–92.

McBride, M.B. (1995) Toxic metal accumulation from agricultural use of sludge: are US EPA regulation protective, *Journal of Environmental Quality* **24**, 5–18.

McGrath, S.P. (1984) Metal concentrations in sludges and soil from a long-term field trial. *Journal of Agricultural Science (Camb.)* **103**, 25–35.

McGrath, S.P. and Cegarra, J. (1992) Chemical extractability of heavy metals during and after long-term sludge applications of sewage sludge to soil. *Journal of Soil Science* **43**, 313–321.

McLaughlin, M.J., Tiller, K.G, Naidu, R. and Stevens, D.P. (1996) Review: the behaviour and environmental impact of contaminants in fertilizers. *Australian Journal of Soil Research* **34**, 1–54.

McLaughlin, M.J., Hamon, R.E., McLaren, R.G., Speir, T.W. and Rogers, S.L. (2000) Review: A bioavailability-based rationale for controlling metal and metalloid contamination of agricultural land in Australia and New Zealand. *Australian Journal of Soil Research* **38**, 1037–1086.

Merrington, G. and Madden, C. (2000) Changes in Cd and Zn Phytoavailability in an agricultural soil after amendment with papermill sludge and biosolids. *Communications in Soil Science and Plant Analysis* **31**, 759–776.

Moolenaar, S.W., Lexmond, T.M. and Van-der-Zee, S.E.A.T.M. (1997) Calculating heavy metal accumulation in soil: a comparison of methods illustrated by a case-study on compost application. *Agriculture, Ecosystems and Environment* **1**, 71–82.

Mullins, G.L. and Mitchell, C.C. (1995) Crops. In: *Soil Amendments; Impacts on Biotic Systems* (Ed. J.E. Rechcigl). CRC Pres, Inc, Boca Raton, Fla, pp. 1–40.

Nemerow, N.L. (1991) *Stream, Lake, Estuary and Ocean Pollution*. Environmental Engineering Series, Van Nostrand Reinhold, New York.

Nicholson, R.J. (1994) Systems of Storage and Disposal of Livestock Wastes. In: *Pollution in Livestock Production* (Eds I. Ap Dewi, R.F.E., Axford, I., Marai, M. Fayez and H. Omed), CAB International, Wallingford, pp. 385–402.

Nicholson, F.A., Hutchinson, M.L., Smith, K.A., Keevil, C.W., Chamber, B.J. and Moore, A. (2000) *A Study on Farm Manure Applications to Agricultural Land and an Assessment of the Risks of Pathogen Transfer into the Food Chain*. Ministry of Agriculture, Food and Fisheries, London.

NRA (1992) *The Influence of Agriculture on the Quality of Natural Waters in England And Wales*. Water Quality Series No. 6, National Rivers Authority, Bristol.

NRA (1995) *Water Pollution Incidents in England and Wales*. Water Quality Series No. 25, National Rivers Authority, Bristol.

O'Donnell, C., Dodd, V.A., Kiely, P.O. and Richardson, M. (1995) A study on the effects of silage effluent on concrete: Part 1, significance of concrete characteristics. *Journal of Agricultural Engineering Research* **60**, 83–92.

O'Donnell, C., Williams, A.G. and Biddlestone, A.J. (1997) The effects of temperature on the effluent production potential of grass silage. *Grass and Forage Science* **52**, 343–349.

Owen, L. (1998) Use of Organic Manure; Storage of Liquid Fertilizers. The British Survey of Fertilizer Practice, 1994, Edinburgh University Data Library, pp. 18.

Pathak, H., Joshi, H.C., Chaudhary, A., Chaudhary, R., Kalra, N. and Dwiwedi, M.K. (1999) Soil amendment with distillery effluent for wheat and rice cultivation. *Water, Air, and Soil Pollution* **113**, 133–140.

Patterson, D.C. and Steen, R.W.J. (1994) The Use of Silage Effluent as an Animal Food. In: *Pollution in Livestock Production Systems* (Eds I. Ap Dewi, R.F.E. Axford, I. Marai, M. Fayez and H. Omed). CAB International, Wallingford, pp. 275–307.

RCEP (1996) Sustainable Use of Soil. Royal Commission On Environmental Pollution 19th Report (Cm 3165). Her Majesty's Stationery Office, London.

Renner, R. (2000) Sewage sludge: pros and cons. *Environmental Science and Technology* October 1, 430A–435A.

Reuter, D.J. and Robinson, J.B. (1997) *Plant Analysis, An Interpretation Manual*. 2nd Edition. CSIRO Publishing, Collingwood, Australia.

Richardson, M., Dodd, V.A., Lenehan, J.J., Conaty, S. and O'Kiely, P. (1999) The influence of cement content and water/cement ratio on the durability of Portland cement concretes exposed to silage effluent. *Journal of Agricultural Engineering Research* **72**, 137–143.

Rund, T. (1995) Heavy metals – finding the status quo, *Arable Farming* (May issue), 30–34.

Schechtner, G. (1992) Prospects of success for aeration of slurry in pasture land. *Veroffentlichungen Bundesanstalt fur Alpenlandische Landwirstschaft Gumpenstein* **16**, 1–82.

Schmidt, J.P. (1997) Understanding phytotoxicity thresholds for trace elements in land-applied sewage sludge. *Journal of Environmental Quality* **26**, 4–10.

Schofield, K., Seager, J. and Merriman, R.P. (1993) The impact of intensive dairy farming activities on river quality: the Eastern Cleddau catchment study. In: *Agriculture and the Environment* (Ed. J.G. Jones). Ellis Horwood Series in Environmental Management, Science and Technology, London, pp.75–90.

Scottish Envirotec (1995) Producing a suitable sludge product for agriculture, *Scottish Envirotec* (June issue), 24–36.

SEPA (1998) *Strategic Review of Organic Waste Spread on Land*. Scottish Environment Protection Agency, Stirling.

Shepherd, M. and Gibbs, P. (2001) *Managing Manure On Organic Farms*. ADAS and Elm Farm Research Centre, Department of Environment, Food and Rural Affairs, London.

Smith, S.R. (1996) *Agricultural Recycling of Sewage Sludge and the Environment*. CAB International, Wallingford.

Smith, K.A. and Chambers, B.J. (1993) Utilizing the nitrogen content of organic manures on farms – problems and practical solutions. *Soil Use and Management* **9**, 105–112.

Smith, K.A. and Chambers, B.J. (1995) Muck: from waste to resource. Utilisation: the inputs and implications. *Agricultural Engineer* **50**, 33–38.

Smith, R.V., Burns, L.C., Doyle, R.M., Lennox, S.D., Kelso, B.H.L., Foy, R.H. and Stevens, R.J. (1997) Free ammonia inhibition of nitrification in river sediments leading to nitrite accumulation. *Journal of Environmental Quality* **26**, 1049–1055.

Smith, K.A., Brewer, A.J., Dauven, A. and Wilson, D.W. (2000) A survey of the production and use of animal manures in England and Wales. I. Pig Manure. *Soil Use and Management* **16**, 124–132.

Smith, K.A., Brewer, A.J., Crabb, J. and Dauven, A. (2001) A survey of the production and use of animal manures in England and Wales. II. Poultry manure. *Soil Use and Management* **17**, 48–56.

Soil Association (1999). *Standards for Organic Food and Farming*. Soil Association, Bristol, p. 113.

Stehouwer, R.C., Wolf, A.M. and Doty, W.T. (2000) Chemical monitoring of sewage sludge in Pennsylvania: variability and application uncertainty. *Journal of Environmental Quality* **29**, 1686–1695.

Stone, K.C., Hunt, P.G., Humenik, F.J. and Johnson, M.H. (1998) Impact of swine waste application on ground and stream water quality in an Eastern Coastal Plain watershed. *Transactions of the ASAE* **41**, 1665–1670.

Tisdale, S.L, Nelson, W.L., Beaton, J.D. and Halvin, J.L. (1993) *Soil Fertility and Fertilizers*. 5th Edition. Prentice Hall, New Jersey.

Tschäpe, H., Prager, R., Streckel, W., Fruth, A., Tietze, E. and Böhme, G. (1995) Verotoxinogenic Citrobacter freundii associated with severe gastroenteritis and cases of haemolytic uraemic syndrome in a nursery school: green butter as the infection source. *Epidemiology and Infection* **114**, 441–450.

US EPA (1993) *Part 503 – Standards for the use and disposal of sewage sludge*. United States Environmental Protection Agency. Federal Register, **58**, 9387–9401.

Van Kessel, J.S., Reeves III, J.B. and Meisinger, J.J. (2000) Nitrogen and carbon mineralization of potential manure components. *Journal of Environmental Quality* **29**, 1669–1677.

Wase, D.A.J. and Thayanithy K. (1994) Biogas production. In: *Pollution in Livestock Production Systems* (Eds I. Ap Dewi, R.F.E. Axford, I. Marai, M. Fayez and H. Omed). CAB International, Wallingford, pp. 333–343.

White, R.E. (1997) *Principles and Practice of Soil Science, the Soil as a Natural Resource*. 3rd Edition. Blackwell Science Ltd, Oxford.

WRc (1996) *Reed Beds and Constructed Wetland for Wastewater Treatment*. WRc, Marlow.

Zibilske, L.M., Clapham, W.M. and Rourke, R.V. (2000) Multiple applications of paper mill sludge in an agricultural system: soil effects. *Journal of Environmental Quality* **29**, 1975–1981.

Gaseous Emissions 6

6.1 INTRODUCTION

Agriculture is both a source and a sink of atmospheric pollution. The effects of agricultural emissions range from the short-term local nuisance of excessive odour, to their contribution to the long-term impact of global warming. This chapter considers the four major polluting gases from agriculture: ammonia, methane, nitrous oxide and carbon dioxide before finally addressing farm odours. Because of their differing behaviour in the atmosphere (Box 6.1), odours and to a certain extent ammonia emissions are seen as local pollution issues, whilst methane, nitrous oxide and carbon dioxide emissions have implications for the global environment. This chapter will address:

- the production and form of the major gaseous emissions from agriculture;
- the potential pollution problems these gases may cause;
- practical solutions to minimising their loss.

Agriculture and gaseous emissions

Natural processes generate gaseous emissions from agriculture. Ammonia, methane, nitrous oxide and carbon dioxide are all components of the natural biogeochemical cycles that circulate carbon and nitrogen between the soil, plants and animals, and the atmosphere. Their significance as agricultural pollutants has only arisen because of the expansion (e.g. increased areas of arable crops), specialisation (e.g. concentration of dairy cows into large herds) and intensification (e.g. increased use of fertiliser nitrogen) of crop and livestock production.

As well as contributing to atmospheric pollution, agriculture is also affected by pollutants emitted from other sources. Nitrogen oxide (NO_x) emissions originating from industrial and transport combustion processes can have a detrimental effect on crop establishment and growth due to their acidifying effect on soils (RCEP, 1996). Full consideration of the detrimental impact of air pollution upon agriculture is beyond the scope of this chapter, but Conway and Pretty (1991) and RCEP (1996) include useful reviews with UK and international examples.

Ironically, one of the main effects of measures to improve air quality and reduce the impact of acid deposition in the UK has been an increase in the incidence of sulphur deficient arable crops, notably oilseed rape and wheat (Rund, 1994). In the 1970s and early 1980s, sulphur dioxide deposition was estimated to account for roughly two-thirds of acid deposition in the UK (RCEP, 1996). While excessive sulphur deposition can damage crops, it is also essential in moderate quantities for crop growth and development. Levels of sulphur deposition have now been reduced to levels of less than a quarter of those in 1980. This is often well below arable crop requirements and farmers are advised to apply 10–30 kg sulphur ha^{-1} $year^{-1}$ to avoid the risk of yield loss (Rund, 1994). This chapter does not consider smoke pollution (MAFF, 1998) or acute occupational exposure to gaseous pollutants, such as ammonia, hydrogen sulphide and nitrogen dioxide, in and around farm buildings.

Box 6.1 The fate of gases in the atmosphere

The earth's atmosphere has never been free of change, but the current rate of alteration is unprecedented and the effects of this are well documented (e.g. acid rain, stratospheric ozone depletion and potential global warming). These phenomena do not stem from changes in the major constituents of the atmosphere (nitrogen and oxygen), but mainly from increases in the levels of several minor or trace gases (MAFF, 2000a).

Change in concentrations of selected atmospheric trace gases (Bunce, 1993).

	1850	1985
Carbon dioxide – CO_2	280 ppmv	345 ppmv
Methane – CH_4	0.6–1.0 ppmv	1.7 ppmv
Nitrous oxide – N_2O	280 ppbv	305 ppbv
CFC-11 + CFC-12	0	0.6 ppbv
Tropospheric ozone – O_3	≈10–50 ppbv	10–50 ppbv

The fate of these gases in the atmosphere varies greatly depending upon their chemical characteristics (Graedel and Crutzen, 1989). Gases which are relatively unreactive and insoluble in water (e.g. methane and nitrous oxide) will rapidly spread through the troposphere (the lower 10–15 km of the atmosphere) and may move into the stratosphere (the next 30–40 km of the atmosphere). Gases which are soluble (e.g. ammonia) are more likely to dissolve in moisture, either on particulate material or in water droplets, and thus be directly and relatively rapidly returned to the earth's surface. They are less likely to diffuse very far through, and above, the troposphere.

A concept used to describe the atmospheric lifetime of gaseous emissions is their residence time. This is calculated as the total amount of gas in the atmosphere divided by its annual rate of removal (Box 6.2).

6.2 AMMONIA EMISSIONS

Pollution problems

Ammonia (NH_3) is a major atmospheric pollutant, but it only has a very short residence time as a gas in the troposphere and consequently plays little role in atmospheric chemistry. It may however, neutralise some of the acidity (H_2SO_4 and HNO_3) (Elsom, 1992) and can also absorb infrared radiation, and its contribution to global warming is considered insignificant (Bouwman, 1990).

Most NH_3 emissions are rapidly returned to the earth's surface via wet (e.g. in mist or rain) or dry deposition, often close to their point of origin. It is estimated that two-thirds of the UK's ammonia emissions are re-deposited within the UK (Fowler *et al.*, 1996; INDITE, 1994). Ammonia emissions have three potential consequences:

- Although NH_3 is alkaline, it contributes to soil and freshwater acidification by the release of H^+ ions during the microbial oxidation (nitrification) of NH_4^+ to NO_3^-.

Since sulphur emissions have declined, nitrogen compounds (including NH_3) now make the greatest contribution to soil and freshwater acidification in the UK and are considered likely to dominate acid deposition in much of Europe in future (RCEP, 1996).

- It raises local levels of soil N, enhancing soil fertility and possibly increasing the risk of losses by leaching an NO_3^-.

- It may have direct toxic effects upon plants causing leaf yellowing and secondary effects such as nutrient imbalance (Van der Erden *et al.,* 1998).

Ammonia re-deposition is a mixed blessing for farmers as it may present a useful supplementary source of N (Chapter 2). However, high levels of localised N input and soil acidification can adversely affect crop yield and quality as was demonstrated by Speirs and Frost (1987) in field crops adjacent to an intensive poultry unit.

Increased NH_3 inputs in sensitive natural and semi-natural ecosystems are a cause of concern because the combination of acidification and N enrichment can significantly change the characteristic composition of flora, reducing both species abundance and biodiversity (Pearson and Stewart, 1993; INDITE, 1994). In freshwater ecosystems, the acidifying effect of NH_3 deposition is likely to be more important than nutrient enrichment (i.e. eutrophication) because N is not generally limiting (INDITE, 1994) (Chapter 3). The sensitivity of natural and semi-natural ecosystems to the impact of atmospheric pollution is variable and is commonly expressed via the concept of critical loads. Critical load is the maximum atmospheric pollutant load that sensitive ecological systems can tolerate without incurring long-term harmful effects (RCEP, 1996). The concept has mainly been applied to acid pollutants to indicate the capacity of soils in sensitive environmental areas to buffer atmospheric acid inputs (CLAG, 1994).

Critical loads for N inputs are less easy to determine, but have been estimated at 5–35 kg N ha^{-1} $year^{-1}$ for a range of ecosystems from heathland to commercial forestry; the poorer the soil and the sparser the vegetation, the smaller the critical load. For acidic coniferous forests the critical load for N may be 7 kg ha^{-1} $year^{-1}$, whereas for calcareous grasslands it may be as high as 35 kg ha^{-1} $year^{-1}$. It is estimated that NH_3 deposition may be responsible for 43% of the area over which critical N loads have been predicted to be exceeded in the UK (Hill, 2000).

Sources of ammonia pollution

The largest source of ammonia emissions in Europe is volatilisation from agriculture (Fowler *et al.,* 1996). In the UK, this may account for more than 80% of the total ammonia emissions (MAFF, 2000a). Other sources include sewage works, industrial combustion, biomass burning and landfill. There are three main circumstances under which NH_3 volatilisation from agriculture occurs:

- From the decomposition of livestock wastes containing urea (cattle, sheep and pigs) or uric acid (poultry): Hydrolysis of the urea or uric acid by the enyzme urease produces NH_3 and CO_2. Since NH_3 is an alkali its production results in a localised increase in pH which shifts the natural equilibrium between NH_3 and NH_4^+ ions in liquid solution towards NH_3. Gaseous ammonia emission then occurs due to the rapid volatilisation of this NH_3 from solution.

- Following applications of ammonium or urea-based fertilisers. The process of emission is similar to that for livestock wastes, with the greatest potential for fertiliser N losses (10–20%) due to NH_3 volatilisation occurring when urea is applied to alkaline soils.

- From the foliage of fertilised crops. Ammonia volatilisation may occur via the stomata of plants throughout the life cycle of a growing crop. The likelihood of this occurring is closely tied to N metabolism during the different phases of crop development, with peaks in emission occurring when plants have excess nitrogen in the form of NH_4^+ in the intercellular solution of their leaves. This is likely both in the periods following fertiliser application, and after anthesis (flowering) when the crop is remobilising N for transfer to the grain by breaking down proteins in the leaves. Other sources of NH_3 emission from fertilised crops include the decomposition of fallen foliage, such as from oilseed rape or where grass has been cut and left in the field.

Nevertheless, it is generally accepted that NH_3 volatilisation is greatest from livestock systems (Jarvis and Pain, 1990) and that emissions have increased significantly with:

- the expansion of the livestock industry (i.e. number of animals);

- the intensification of livestock production, including increased rates of fertiliser N application to grassland and the use of high protein diets for enhanced productivity (only about 20–40% of N fed to livestock as protein in forage or feed is actually retained in the animal or animal products, the remainder is excreted as non-absorbed N in faeces and urea in urine).

Estimates of NH_3 emissions from agriculture remain variable, and are complicated by the different sources of NH_3 and the wide range of factors affecting the rate of emission, including temperature, seasonality, soil pH and farm management practices (e.g. grazing systems, choice of livestock housing system, waste management etc.). Fowler *et al.* (1996) prepared an inventory of NH_3 emissions for the UK which suggested total emissions in the region of 450,000 t NH_3 year^{-1}, of which 406,000 t year^{-1} (over 90%) were from agriculture. These estimates have been further modified, as emission inventories become more refined, Pain *et al.* (1999) suggested 230,000 t year^{-1} and Misselbrook *et al.* (2000) 226,000 t year^{-1}, with 80% coming from agriculture. The majority of losses (40%) come from animal housing where manures often accumulate, followed by spreading manure and slurry to land (30%), livestock grazing (12%), manure and slurry storage (9%) and the use of nitrogenous fertilisers (9%) (MAFF, 1998; Pain *et al.*, 1999).

Practical solutions

Both the European Union and the UN Economic Commission for Europe are concerned with strategies and protocols to combat the effects of nitrogen and acid deposition upon sensitive ecosystems (RCEP, 1996). At present these strategies are concerned mainly with industrial pollution and do not yet address NH_3 emissions, although it is feasible that controls to reduce NH_3 loss will be imposed upon farmers in the future.

In the meantime, efforts to reduce NH_3 pollution remain at the discretion of individual farmers, although there is clearly an economic incentive to reduce losses in order to retain and profitably utilise as much N within the farm system as possible. Jarvis (1993) estimated that NH_3 volatilisation from grazed pasture, animal housing, manure storage and spreading accounted for losses of 3.5 t N $year^{-1}$ on a 76 ha model dairy farm i.e. an average of 46 kg N ha^{-1} $year^{-1}$. Assuming that the majority of this NH_3 was derived from purchased fertiliser N (at approximately £0.30 kg^{-1}), this was equivalent to a direct financial loss of over £1000 $year^{-1}$.

Minimising ammonia emissions by good practice

The main source of advice on minimising ammonia emissions for farmers in the UK is the revised MAFF Code of Good Agricultural Practice for the Protection of Air (MAFF, 1998) which dedicates a complete chapter to NH_3 and reducing NH_3 loss.

The most important strategy for minimising NH_3 emission is to limit the opportunity for volatilisation to occur by reducing the contact of manure and slurry with the air during livestock housing, waste storage and spreading of organic wastes. Methods for controlling odour are also useful for reducing NH_3 losses (see below). For example, alternative slurry application methods such as band spreading and injection can reduce emissions by 40–50% under UK conditions (MAFF, 1997a, 2001a,b). Further specific points of good practice are:

- Where practicable match protein levels in feed more closely with livestock requirements (e.g. according to growth stage and/or the target level of production) in order to reduce residual N in wastes. In feeding trials, pigs fed a modified low crude protein diet excreted 44% less total N in slurry, while NH_3 concentrations in their pig houses were reduced by around 50% (MAFF, 1997b). Growth rates and finished weights were the same as pigs fed a normal commercial diet. Similar techniques have been developed for poultry, but increasing the efficiency of dietary N use by cattle and sheep is more difficult.

- As with odours, the key to low ammonia emissions from livestock housing is cleanliness and keeping the floors and walls clear of urine and faeces. This is greatly aided by maintenance and upkeep of flushing systems and scrapers.

- Slurry storage may only represent about 2% of the total annual NH_3 loss, but recent research suggests that a cheap and relatively straight-forward way of mitigating losses is to float a cover material over the top. Plastic films have been found to be particularly effective on slurry tanks, reducing emissions by as much as 80%. For lagoons, stabilised welded plastic film is able to reduce NH_3 loss by 95%, although venting to remove biogas and prevent the build-up pressure is needed (MAFF, 2001c).

6.3 METHANE EMISSIONS

Pollution problems

Methane (CH_4) is a natural gaseous component of the atmosphere and along with carbon dioxide, nitrous oxide and water vapour is particularly important in regard to the greenhouse effect. Unlike ammonia, the residence time of CH_4 may be considerable, and may absorb 21 times more infrared radiation than CO_2 (Box 6.2). The major polluting action of CH_4 is associated with its role in increased rates of global warming. Importantly CH_4 is odourless, although often wrongly blamed as the cause of malodour. However, as pointed out in the revised air code (MAFF, 1998) in regard to the potential build up of CH_4 in confined spaces such as tanks or storage facilities, it is highly flammable.

Sources

Methane emission from agriculture accounts for almost 37% of the UK total, and is produced primarily by the natural digestive processes of ruminants (MAFF, 2000b). Cattle account for 75% of the CH_4 produced by livestock (1 mature dairy cow may produce 100 kg of CH_4 in a year), followed in order of decreasing importance by sheep, pigs and poultry. Poultry do not directly produce methane, although under certain conditions their manure may do so (MAFF, 1998). Indeed, the action of bacteria on most slurries and manures may produce CH_4, although in comparison to the rate of production by livestock it is much slower and promoted by conditions of anaerobic soil respiration (MAFF, 1998).

Of considerable global importance is the production of CH_4 from flooded rice paddies. Indeed, globally, rice production may be responsible for 30% of all CH_4 emissions. The primary cause is due the anaerobic fermentation of organic matter in flooded soils, two of the end products being CO_2 and CH_4 (Neue, 1997; White, 1997). This same process is responsible for the natural emission of methane from native wetlands, bogs and marshes (Kasimir-Klemedtsson *et al.,* 1997). With a predicted increase in rice production of 65% in the next 25 years, it is clear this source is likely to become much more important and the search for mitigating management strategies more earnest. However, some workers suggest technologies already exist which if implemented, could see decreases in CH_4 emissions from flooded paddy soils of 40% (Minami *et al.,* 1998; Bodeller *et al.,* 2000).

A major non-agricultural source of biogenic CH_4 is refuse landfills, although the output is enormously variable, spanning 7 orders of magnitude (<0.0004–>4000 g m^{-2}). Nevertheless, the contribution of CH_4 from this source is still currently greater than from agriculture. However, this is likely to be reduced as stricter legislative controls on landfills begin to take affect.

Practical solutions

There is limited advice given by MAFF for the reduction of CH_4 emissions from livestock. Reductions in the numbers and intensity of operations would have a significant effect on emissions as would increasing the productivity per animal. Currently, the research focus is upon:

- dietary modifications and attempts to change the protozoan content of the rumen, which plays a major role in the generation of CH_4;
- improved health and welfare of livestock;
- maximising genetic potential.

However, it is thought that in the short-term these will make little difference to CH_4 emissions from agriculture (MAFF, 1998, 2000a).

In terms of CH_4 emissions from manures, reductions have been observed during composting of deep litter from pig production based on the density of the heap (Sommer and Moller, 2000). Furthermore, alternative slurry and manure handling and storage facilities along with the promotion of the use of the generated CH_4 (as biogas) as an alternative energy source may also be emission mitigating possibilities (MAFF, 2000a).

Box 6.2 Greenhouse gases and global warming

The greenhouse effect is an essential and natural feature of the earth's atmosphere. If there were no greenhouse effect the surface of the earth would be cooler by almost 30 °C. The absorption of outgoing long-wave radiation reflected from the earth's surface by certain atmospheric gases reduces the amount of heat energy that would otherwise be lost to space, so behaving akin to a thermal blanket (Bunce, 1993; MAFF, 2000a). An increase in the concentration of these atmospheric gases responsible for this greenhouse effect may further increase the amount of heat energy trapped. Therefore, the anthropogenic influence upon the greenhouse effect is thought to be primarily associated with increased global warming due to elevated emissions of greenhouse gases. The past decade in the UK has been the warmest in over 300 years (MAFF, 2000a). This evidence, coupled with measures of increases in greenhouse gases (Box 6.1) and the results of predictive climate models have been cited by the Intergovernmental Panel on Climate Change as evidence of human induced changes to our climate (Young, 2000).

Agriculture is responsible for around 8% of the greenhouse gas emissions in the UK (MAFF, 2000a) although globally it is thought that 20% of the CO_2, 35% of the CH_4 and 65% of the N_2O come from soil processes or as a result of land-use change (Smith, 1997).

The establishment of a national inventory of greenhouse gases is the first step in being able to establish and target reductions in gas emissions from agriculture and other sources and is required by the United Nations Framework Convention on Climate Change (DETR, 1998; MAFF, 2001c). Even if a reduction in emissions occurs immediately, as can be seen from the table, the likely beneficial effects are unlikely to be seen in the short to medium term.

Estimates from global climate models suggest that by 2050 the UK will be 1–2 °C warmer than the 1961–1990 average, with winters warmer and hot spells during summers increasing, although changes in rainfall are difficult to predict (MAFF, 2000a).

Potential impacts of atmospheric gases upon global warming (MAFF, 2000a)

	Residence time (years)	Relative impact	Contribution (%)
CO_2	2–230	1	60
CH_4	10	25	15
N_2O	150	200	5

The likely impacts upon agriculture include:

- change in the location of agricultural activities (the range of current crops may move further north);
- new crop varieties previously unsuitable for climatic conditions in the UK will need to be used;
- water availability limitations with a large impact on high value horticultural crops;
- yields of some crops and grasses may increase, and so may weeds, with implications for herbicide resistance;
- changes in the types of pests and diseases currently affecting crops and livestock, with implications for the transmission of vector borne diseases.

The adaptability of UK agriculture stands it in good stead to assimilate and accommodate potential changes forecast by the current predicted climatic models (MAFF, 2000a). However, some commentators suggest that the evidence currently presented to support increased global warming due to anthropogenic induced climate change is flawed, or at least not as irrevocable as some scientists are suggesting (Young, 2000). In particular, there is the concern that not all evidence supports global warming. While surface temperatures have risen in the northern-hemisphere they have fallen in the southern hemisphere. Furthermore, it is difficult to differentiate between natural climatic variation and human induced change. Temperatures have risen and fallen dramatically in the past by at least 2 °C, allowing; the Vikings to settle Greenland between the ninth and twelfth centuries, and producing the Little Ice Age in the late seventeenth century. It is also well known that many of the climatic models currently being relied upon to produce scenarios of change are rarely able to adequately consider dust emissions or natural changes in CO_2 levels (Young, 2000). Nevertheless, increased global warming would have such an extensive effect upon so many facets of our lives that a precautionary approach in regard to emission controls of greenhouse gases seems the wisest option.

6.4 NITROUS OXIDE EMISSIONS

Pollution problems

The importance of nitrous oxide (N_2O) as a potential gaseous pollutant was identified in Chapter 2. The impacts of N_2O are two-fold behaving as a greenhouse gas (Box 6.2) and stratospheric ozone depleter, further degrading climatic conditions with potential influences upon agriculture (Colbeck and Mackenzie, 1996). The global warming potential of N_2O, is 200–300 times greater relative to CO_2 (Box 6.2) (Schlesinger, 1997).

Sources of pollution

The release of N_2O resulting from agricultural activities including the stimulation of soil bacteria, manures and slurries accounts for over 50% of all emissions in the UK (MAFF, 2000b). Soil is thought to be responsible for one quarter of all global N_2O emissions (Mosier *et al.,* 1998). Other significant sources of N_2O – N include emissions from animals as well as housing and waste storage (17%), from leached N (21%) and from NH_3 (3%) (MAFF, 2001d). Indirect emissions of N_2O can come from atmospheric N deposition, leached nitrate and sewage treatment plants, although their magnitude is yet to be satisfactorily gauged. The primary mechanism thought to be responsible for the loss of N_2O and N_2 from soil is denitrification under anaerobic conditions in which the nitrate and nitrite are reduced (Rowell, 1994):

NO_2^- and	NO_3^- →	$N_2O\uparrow$ →	$N_2\uparrow$
Nitrite and	Nitrate	Nitrous oxide gas	Dinitrogen gas

A small amount of NH_4-N is converted to N_2O and NO during nitrification. These mechanisms are natural and it is their enhancement through anthropogenic actions that is of environmental importance. The production of N_2 is thought to be of little concern environmentally, although clearly economically undesirable. As much as 30%, although more commonly between 0.5 and 2.0% of applied fertiliser N is lost as N_2O (Tisdale *et al.,* 1993; Smith *et al.,* 1997; Baggs *et al.,* 2000). Not surprisingly the more N that is cycled through the agroecosystem, through the increased use of inorganic N fertiliser and intensification of arable and livestock production systems, the greater the enhancement and the larger the quantity converted to gaseous nitrogen forms, including N_2O (Smith *et al.,* 1997).

The exact proportion of N_2O or N_2 produced during denitrification is difficult to ascertain, but it has been suggested that N_2O emissions may be more important under less anaerobic conditions. While only a few facultative aerobic bacteria (species belonging to the genera *Pseudomonas*, *Bacillus* and *Paracoccus*) and several autotrophs (*Thiobacillus denitrificans*, *T. thioparus*) are responsible for denitrification there tends to be large populations of these organisms in arable soils (Tisdale *et al.,* 1993). There are a range of interrelated factors that may promote the gaseous loss of N from soils and depletion of oxygen through the creation of anaerobism including:

- soil moisture content;
- the amount and nature of organic matter present (readily decomposable C substrates can stimulate significant denitrification);
- aeration (which can be influenced greatly by farm traffic and livestock);
- soil pH;
- soil temperature (denitrification tends to be limited at a pH <6 and temperature <0 °C);
- the form and timing of fertiliser N application (Tisdale *et al.,* 1993).

In a study on the effects of irrigation, crop management and N fertilisation on N_2O loss it was observed that the greatest emissions occurred in treatments which received irrigation water, fertiliser N and remained fallow (Simojoki and Jaakkola, 2000). When water filled 60–90% of soil pore space N_2O emissions doubled compared with unirrigated controls. If predicted climate changes do occur (Box 6.3) and fertiliser use and irrigation also increase as predicted, then it is likely that emissions of N_2O are set to rise dramatically (Smith *et al.,* 1997). It is also suggested that there is a strong correlation between N_2O emission and soil mineral N content which may be greatly influenced by the quantity of applied N fertiliser (Smith *et al.,* 1997; MAFF, 2000a).

There appears to be less conclusive evidence of a relationship between the form of N fertiliser addition and N_2O emissions. Although Clayton *et al.* (1997) noted that there were significant differences in gaseous losses from grassland cut for silage having been fertilised with 3 applications of different N sources at a rate of 120 kg N ha^{-1} (Figure 6.1). Nevertheless, the effect of inorganic N fertiliser type is thought to be small in comparison with other environmental effects (Barton *et al.,* 2000).

The influence of the incorporation of plant residues and manures on the emission of N_2O may be considerable. With the magnitude depending largely upon the residue composition and the amount incorporated, as well as factors listed above including soil type and land management. In New Zealand, over 50% of the anthropogenic N_2O emissions are from livestock excrement (De Klein *et al.,* 2001). Generally, emissions are greatest when residues having low C:N ratios are incorporated into soil (Chapters 2 and 5), and in some circumstances may also be elevated following deep ploughing, possibly due to increased microbial access to labile C in the form of degrading root material. However, Baggs *et al.* (2000) noted that a return to background N_2O emissions occurred after approximately two weeks, possibly due to the decrease in easily degradable C substrate. Scott *et al.* (2000) also observed peaks in N_2O production following the incorporation of sewage sludge into grassland soils and noted further peaks in emissions after rainfall events (as is often observed after irrigation, De Klein *et al.,* 1999). This work suggested that the losses of N_2O from grasslands amended with sewage sludge might be as significant as those receiving inorganic N fertiliser.

Intensively managed grassland systems may be considerable sources of N_2O emission in many parts of the UK and Europe, primarily due to the large N input and the possibility of wet soil conditions (Velthof and Oenema, 1995). This is exacerbated by the concentration of herbage N in urine and manure and by the compaction of the soil due to trampling. The amount of N_2O released expressed as a percentage of N excreted by livestock is between 0.2 and 9.9% (Chapter 2) (Oenema *et al.,* 1997). This considerable variability is a reflection of the complexity of making measures of N_2O and stresses the multifaceted nature of this problem.

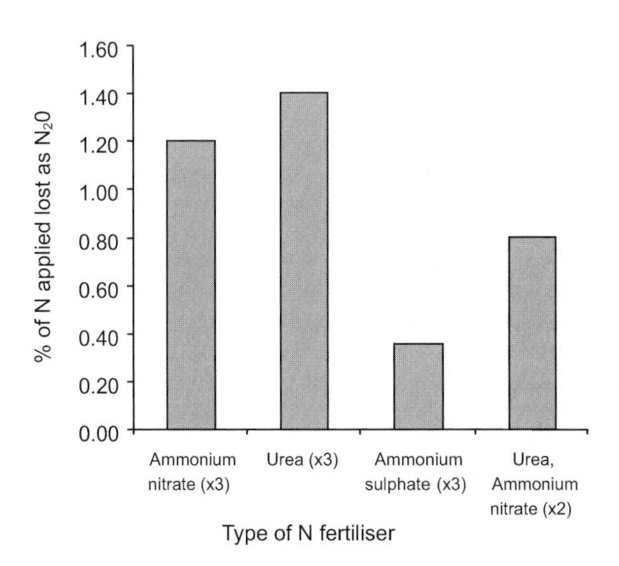

Figure 6.1 The cumulative N_2O emissions from grassland cut for silage fertilised with different sources of N (Clayton *et al.,* 1997).

Practical solutions

The complex relationship that exist between the environment and land management factors influencing N_2O release would suggest that practical, mitigating solutions may be few and far between. However, strategies outlined in Chapter 2 to minimise N loss and also in Chapter 3 to reduce P loss are likely to be beneficial in reducing gaseous emissions (MAFF, 1998, 2000a). Strategies include (Cole *et al.,* 1996; MAFF, 2000a):

- matching N supply with crop demand through use of N testing of plants and split applications;
- reducing farm N inputs and increasing N efficiency, utilising crop residues and accounting for N inputs from manures;
- utilising better fertiliser techniques – placement, nitrification inhibitors, foliar sprays and slow release sources;
- taking account of soil water content and drainage. Water-filled pore space should be <60% and their should be reduced traffic and tillage operations as the creation of

zones of reduce air and water permeability may produce localised zones of anaerobism (Smith *et al.*, 1997).

Management options to lower the loss of N_2O from grazed grasslands include reducing stocking density, reducing N content of urine and decreasing the number of manure and dung patches. Suggested methodologies to achieve this may involve increasing the productivity per cow, the use of low N feeds, less inorganic N applications and delayed and restricted grazing (Oenema *et al.*, 1997).

In conclusion, the following of good agricultural practice in regard to fertiliser management and the use modern techniques of application coupled with optimisation of irrigation and tillage are likely to be effective in reducing N_2O release. However, these reductions are thought to be insignificant in comparison to the likely increases brought about by the predicted increase in worldwide N fertiliser use (Smith *et al.*, 1997).

6.5 CARBON DIOXIDE EMISSIONS

Pollution problems

The global C cycle and the relevance of carbon dioxide (CO_2) and CH_4 to this cycle are shown in Figure 6.2. The carbon cycle is highly dynamic and it is often stated that rather than one cycle, a number of separate cycles operate on different time scales (Kasting, 2001).

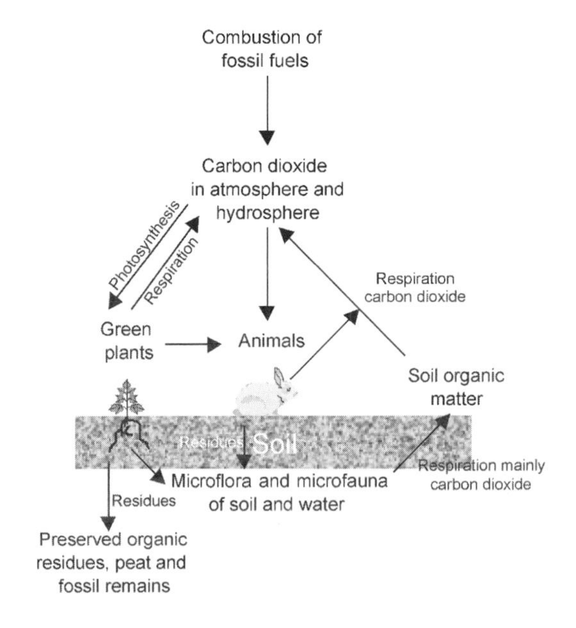

Figure 6.2 The carbon cycle (White, 1997).

Carbon dioxide is the most important of the greenhouse gases, but due to its solubility in water considerable exchange occurs with the worlds oceans which limits its atmospheric lifetime to approximately 2 years (Schlesinger, 1997). The change in atmospheric concentrations of CO_2 are attributed to an increase in source strengths, such as the combustion of fossil fuels, and decreased sink strengths resulting from deforestation and a reduced global rate of photosynthesis (Bunce, 1993). Soils play a particularly important role as a sink for CO_2 acting as a transient store or repository for carbon. Agricultural practices that reduce soil carbon levels, by degrading organic matter therefore reduce the capacity of this sink for C (Table 6.1).

Sources of pollution

In the UK, agricultural production of CO_2 through fossil fuel combustion and lime use accounts for less than 1% of the total released into the atmosphere each year (MAFF, 1998). This may therefore give the impression that agriculture is unimportant in relation to emissions of CO_2. However, there is more carbon in the worlds soils than in plant biomass and the atmosphere put together. It is not surprising therefore that the effects of land-use change induced by the drive towards greater production with the resulting elevated turnover of biomass and commensurate increase in CO_2 may be responsible for 5% of the total UK CO_2 emissions (MAFF, 2000a). Of equal importance is that this loss of soil C to the atmosphere also indicates widespread soil degradation (Grant *et al.*, 2001). An indication of the potential effects of cultivation on soil C is shown in Table 6.1.

Table 6.1 Changes in soil C and N brought about by different cultivation practice in a permanent rotation trial over a 75-year period (Grace *et al.*, 1995).

Rotation	C ($g kg^{-1}$)	N ($g kg^{-1}$)
Permanent pasture	28.8	2.35
Pasture 4 years, wheat 2 years	20.8	1.50
Continuous wheat	16.3	1.15
Wheat, fallow	10.6	0.77
Uncultivated soil	≈20	1.30

Photosynthesis is a transient sink for CO_2, as atmospheric removal is generally balanced over the year by subsequent release after harvesting and so is not a net consumer of CO_2 (MAFF, 2000a).

Practical solutions

Under the auspices of the Kyoto Protocol, the European Union is committed to reducing CO_2 emissions, with the UK looking to reduce 1990 emissions by 12.5% by 2008–2012 (Smith *et al.*, 2000). Strategies for reducing emissions of CO_2 from fossil fuel combustion include energy saving measures such as, ensuring machinery in running optimally, reducing heat loss from buildings, utilising non-fossil fuel sources of energy and precisely controlling temperatures of glasshouses and polythene covered structures. A further indirect source of emissions is through the manufacturer of nitrogenous fertilisers; therefore management practice that promotes efficient N usage will also limit CO_2 losses (MAFF, 2000a).

In terms of CO_2 release due to soil cultivation and management practice, no-tillage systems, avoiding the cultivation of permanent pasture and drainage and cultivation of peatlands have all been observed to greatly reduce loss (Robertson *et al.,* 2000). Soils may also be considered as a C sink, and due to the fact that many agricultural soils are now relatively C-depleted they represent a potential sink for CO_2 if the C lost can be regained (Paustian *et al.,* 1997). Under Article 3.4 of the Kyoto Protocol carbon mitigation and sequestration land-management strategies were encouraged. Potential strategies and scenarios that may provide environmentally and economically viable methods for the improved management of agricultural soils in regard to the removal atmospheric CO_2 include; the increase in area to which animal manures and sewage sludge are applied, straw incorporation, no tillage systems (due less consumption of fossil fuels), agricultural extensification, natural woodland regeneration, bioenergy crop production (such as willows). The combination of these management options offer greater potential for the reduction of agricultural CO_2 emissions and also greater sequestration of CO_2 from other sources (Robertson *et al.,* 2000; Smith *et al.,* 2000: Grant *et al.,* 2001). Even so, the C sequestration strategies would, if carried out extensively across the UK, still only sequester between 6 and 7% of the total CO_2 emissions from all sources (Smith *et al.,* 2000).

6.6 FARM ODOURS

Sources of pollution

Odours are a very complex comprising a number of important properties. While some odours may be considered as pleasant, they are generally regarded as an unhealthy annoyance that should be avoided. As a pollutant, odours cause harm by affecting the well-being of people by eliciting unpleasant sensations and possibly triggering harmful reactions, such as nausea, vomiting and headaches (Elsom, 1992). Some salient properties of odours include:

- intensity – the strength of the odour sensation caused by the concentration of odour substances present;
- odour character – properties which enable substances to be detected;
- hedonic tone – scale of pleasantness or unpleasantness;
- frequency/duration of odour releases.

In the UK, local authority Environmental Health Departments are responsible for enforcing legislation on odour nuisance (MAFF, 1998). Under Part III, Section 79, of the Environmental Protection Act 1990, local authorities are required both to detect statutory nuisances and to conduct investigations. These statutory nuisances include any smell arising on trade or businesses premises, or any animal kept in such a place or manner as to be prejudicial to health or a nuisance.

Agricultural odours are principally associated with livestock farming and are mainly due to the anaerobic breakdown of plant fibre and protein in faeces and urine. They are a form of point source pollution (i.e. with an easily identifiable origin) and are caused complex mixtures of a large number of chemical compounds (e.g. over 77 in pig slurry), including volatile fatty acids, organic acids, phenols, amines and organo-sulphides. The most unpleasant odours tend to be from those substances containing sulphur and nitrogen (Elsom, 1992). The most common sources of farm odour are:

- farm buildings housing animals, such as intensive poultry units;
- slurry or manure stores;
- spreading of slurry or manure;
- silage clamps and other types of animal feed.

A number of factors influence the risk of odour problems occurring from livestock enterprises. These include:

- the distance from neighbouring properties and the local topography;
- the number and type of livestock;
- the prevailing wind direction in relation to neighbouring properties;
- the management of the livestock and housing system used;
- the type and size of slurry store and the way in which it is managed;
- the type of feed used.

Between 1990 and 1996 there were 9000 complaints reported by local authorities and recorded by environmental health officers, involving over 3500 farm premises (MAFF, 1998). Of these complaints the majority were in regard to the spreading of manure and slurry. Pigs were the most common livestock enterprise complained about (50%), then poultry units (30%) and cattle farms (20%), respectively (MAFF, 1998).

Generally, the highest odour emissions occur while the spreading of waste is actually taking place, although it is possible for emissions over the next 8–12 hours to be enough to cause a nuisance (MAFF, 1998). The main reason for high odour levels during spreading is the design of the conventional slurry spreader. This forces a jet of slurry under pressure against a 'splash-plate' causing it to shatter into small droplets which are distributed behind the tanker as a 'spray' (Figure 6.3). Depending upon the trajectory of the 'spray' and the size of the droplets, this greatly increases the release of volatile compounds into the air (especially if the spreader produces a high 'spray' trajectory and small droplets) causing localised odour concentrations up to 15 times greater during spreading than immediately afterwards. Other factors affecting the amount of odour emitted during or after spreading include:

- the rate of application;
- the type of livestock waste, with pig slurry often the most malodorous;
- whether the waste contains milk or silage effluent since these tend to increase the amount of odour released;
- the method and length of storage.

Figure 6.3 Spreading slurry using a vacuum tanker fitted with conventional 'splash-plate' spreading mechanism.

The weather also plays a key role in determining if odour will be a nuisance. The least suitable atmospheric conditions for spreading are high humidities and very light winds since this prevents odours from dispersing. The proximity of spreading operations to houses and the direction of the wind will also affect the likelihood of causing nuisance.

Practical solutions

Comprehensive practical guidance on avoiding air pollution from odours is provided to UK farmers in the MAFF publication, Code of Good Agricultural Practice for the Protection of Air (1998) and much of the material below is derived from this. It is not a statutory code and following it will not provide a defence against a legal charge of causing air pollution.

Furthermore, awareness of farmers of this code is very limited, with just over 20% aware (and <10% owning an actual copy), despite its re-launch in 1998 (MAFF, 2000b).

Whilst it is not possible to avoid all odours from agriculture, the risk of nuisance can be minimised. The first line of defence is to ensure, as far as possible, that all potential sources are a reasonable distance from potential complainants to allow for maximum dispersal and dilution of the odour. This has implications especially for the location of new livestock units and the selection of fields for spreading waste. However, reliance upon the dispersal and dilution of odours to avoid complaints is not always sufficient and attempts should be made to address the causes of emission at source by applying good design and management practices. These include:

- housing and feeding of cattle, pigs and poultry;
- storage and handling of livestock wastes, and;
- spreading of slurry and manure.

In extreme cases, such as the particularly close proximity of existing livestock units to housing, the farmer may need to go beyond good practice and use special equipment and techniques for more rigorous odour control.

Livestock dietary modification and manipulation to reduce the production of odorous compounds in manure have focussed firstly on increased nutrient utilisation (Chapter 3), secondly, efforts to enhance microbial metabolism in the lower digestive tract to reduce excretion of odour causing compounds and finally on changing the physical and chemical characteristics of urine and faeces. Through the use of synthetic amino acids with reduced protein levels significant reductions in odour production (and N) have been observed in pig manures (Sutton *et al.*, 1999). Importantly, it is also thought that such dietary modifications are unlikely to present any major economic burden to the farmer and therefore could be readily adopted.

The location and design of new livestock units

Complaints about farm odour are most frequent when livestock units are located close to housing or other centres of public activity. Local planning authorities therefore have a role to play in avoiding odour nuisance from new livestock units. In the UK, this involves three regulatory measures:

- Under the Town and Country Planning (General Permitted Development) Order 1995, planning permission is required for new facilities, such as livestock buildings and slurry stores, which are situated within 400 m of a residential development. This rule was specifically introduced to reduce the risk of odour nuisance from new or expanding livestock units.
- For new livestock facilities which do not require planning permission, a prior notification system is in place which means that the local planning authority must be informed of the development. The planning authority will then decide within 28 days whether it wishes to approve details of any aspect of the proposed development.

- The Town and Country Planning (Assessment of Environmental Effects) Regulations 1988 require an environmental assessment to be carried out for certain types of development project likely to have significant effects on the environment. In the case of agriculture, this is likely to include new pig units with more than 400 sows or 5000 fattening pigs, and new poultry units with more than 100,000 broilers or 50,000 layers (MAFF, 1998).

The relatively recent EU Integrated Pollution Prevention and Control (IPPC) Directive affects the agricultural industry by introducing controls over emissions to the atmosphere from agricultural activities (Robinson, 1999). For example, IPPC emphasises best available technology and minimising environmental emissions, applying to livestock production units with more than 40,000 birds, 750 sows or 2000 weaners. New livestock housing (e.g. broiler houses) should also be designed with odour control in mind. This includes:

- ensuring adequate ventilation to avoid the development of humid conditions which give rise to unpleasant odours, high levels of ammonia and poor animal health;

- positioning ventilation outlets as high as possible for good odour dispersal and to avoid the strong odours that occur at floor level;

- providing adequate space allowances for livestock, well-designed floors (e.g. self-cleaning slats) and effective waste collection systems to ensure that animals do not get excessively dirty;

- piping or channelling wastes and waste water to storage facilities;

- ensuring drinkers are designed and located to minimise spillage and thus help keep bedding as dry as possible.

Minimising odour nuisance by good management practice

The general points of good practice for the day-to-day management of livestock housing and waste disposal to minimise the risk of causing odour nuisance are summarised in Table 6.2. For full details refer to the Code of Good Agricultural Practice for the Protection of Air (MAFF, 1998).

Table 6.2 General points of good practice for reducing odour emissions and the risk of nuisance from livestock production systems (summarised from MAFF, 1998).

Management of livestock housing and feeding
Maintain a high degree of hygiene and cleanliness.
Clean livestock buildings regularly. Whenever possible, collect and transfer slurry and dung from non-bedded concrete areas to a suitable store every day. If livestock are produced in batches, thoroughly clean and disinfect buildings after each batch of stock is removed.
Where bedding is used, ensure sufficient is used to keep animals clean.
Maintain drains and repair broken or badly laid concrete to prevent effluents from ponding within buildings.

Clean out grit and sediment from slurry channels, collection systems and stores to avoid the development of sediments.
Manage drinking systems to avoid overflow and spillage.
Keep concrete areas around buildings clean and free from build-up of slurry or manure.

Storing and handling manure and slurry

To minimise odour from solid manures encourage natural composting by helping air to penetrate the manure.
Do not add feeds such as milk, whey or silage effluent to slurry. Where practicable, silage effluent should be contained in a separate collection tank.
If slurry is frequently agitated, choose weather conditions when the dispersal and dilution of odours will be greatest i.e. sunny, windy days.
Keep poultry manure as dry as possible to minimise odour problems. If practicable, manure should be store under cover on an impermeable base until it is spread on the land. If field heaps are used, they should be as far away as practical from housing.

Precautions when spreading manure and slurry

Use a weather forecast to help choose suitable conditions for spreading when dispersal and dilution of odours will be greatest i.e. sunny, windy days.
Avoid spreading at weekends, bank holidays or in the evenings unless it is absolutely necessary.
Avoid spreading in fields close to or upwind of houses if using a conventional 'splash-plate' spreader.
Avoid applying more than 50 m^3 of slurry or 50 tonnes of manure per hectare at one time if odour could be a problem.
Where practicable, lightly cultivate or plough immediately after surface spreading to incorporate the slurry or manure as soon as possible.
If there is a risk of causing odour nuisance from slurry irrigation systems, choose low pressure systems producing a low trajectory and large droplets.

Technologies for reducing odour nuisance

In sensitive situations there are various technological options for odour control that go beyond the reductions achievable with good management practice. For example (Culpin, 1992; MAFF, 1998):

- Odours from livestock buildings can be treated by passing the air expelled from ventilators through an air cleaning device or biofilter system. This is a very expensive option restricted to extreme cases of nuisance.
- Odorous emissions during storage and spreading of slurry can be reduced by up to 90% by mechanical separation followed by aerobic treatment to encourage bacterial activity. There are a range of mechanical separators available which remove course solids (10–20% of volume) from slurry leaving a liquid which flows easily and can be pumped to a tank for aeration. The solid fraction left is relatively mild smelling and can be stored and spread like farmyard manure. Aeration methods vary greatly, but the general principle is to produce a stream of small bubbles that maximises oxygen transfer to the liquid. Some of the most successful aerators are electrically-operated 'venturi agitators' which force air downwards into, and around, the storage tank. Increasing the aeration period will allow the slurry to be stored for at least a month or more before the odour returns.

- Emissions during spreading can also be reduced by using an alternative design of slurry spreader. If the fields in which slurry is being spread are a reasonable distance from houses, a conventional 'splash-plate' spreader can be used (preferably one that produces a low 'spray' trajectory and large slurry droplets).

(a)

(b)

Figure 6.4 Alternatives to conventional slurry spreaders include: (a) slurry injectors, and (b) band spreaders (MAFF, 1998).

For more rigorous odour control, band spreaders (also called dribble bar applicators) or injectors can be used (Culpin, 1992; MAFF, 1998; Figure 6.4). Band spreaders discharge slurry directly onto the soil surface through a series of trailing pipes, while injectors apply the slurry below the soil surface. Shallow injection discharges slurry into a series of grooves cut 50–80 mm deep in the soil surface and then closes the grooves with press wheels or discs. Deep injection uses winged injector tines to apply slurry at a depth of 120–300 mm and ensure thorough mixing with the soil. Although more expensive to purchase and to run (e.g. deep injection requires a significantly more powerful tractor), these alternative spreader designs can reduce odour emissions significantly. According to MAFF (1998) when compared to 'splash-plate' spreaders, band spreading can reduce odour by 55–60% and injection by about 85%.

REFERENCES

Baggs, E.M., Rees, R.M., Smith, K.A. and Vinten, A.J.A. (2000) Nitrous oxide emission from soils after incorporating crop residues. *Soil Use and Management* **16**, 82–87.

Barton, L., Roberts, A.C., Vojvodić-Vuković, M. and Rh. Lloyd-Jones, A. (2000) Nitrous oxide emissions from soils irrigated with dairy farm effluent. In: *Soil 2000: New Horizons For A New Century. Australian And New Zealand Second Joint Soils Conference Volume 2: Oral Papers* (Eds J.A. Adams and A.K. Metherell). 3–8 December 2000, Lincoln University, New Zealand Society of Soil Science, pp. 15–16.

Bodeller, P.L.E., Roslev, P., Henckel, T. and Frenzel, P. (2000) Stimulation by ammonium-based fertilisers of methane oxidation in soil around rice roots. *Nature* **403**, 421–424.

Bouwman, A.F. (1990) Exchange of Greenhouse Gases between Terrestrial Ecosystems and the Atmosphere. In: *Soils And The Greenhouse Effect* (Ed. A.F. Bouwman). John Wiley & Sons, Chichester, pp. 61–127.

Bunce, N.J. (1993) *Introduction to Environmental Chemistry*. Wuerz Publishing Ltd, Winnipeg, Canada.

CLAG (1994) *Critical Loads for Acidity in the United Kingdom*. Critical Loads Advisory Group Summary Report. Institute of Terrestrial Ecology, Edinburgh.

Clayton, H., McTaggart, I.P., Parker, J., Swan, L. and Smith, K.A. (1997) Nitrous oxide emissions from fertilized grassland: a two-year study of the effects of N fertilizer form and environmental conditions. *Biology and Fertility of Soils* **25**, 252–260.

Colbeck, I. and Mackenzie, A.R. (1996) Chemistry and pollution of the stratosphere. In: *Pollution, Causes, Effects and Control* (Ed. R.M. Harrison). 3rd Edition. The Royal Society of Chemistry, Cambridge, pp. 194–220.

Cole, V., Cerri, C., Minami, K., Mosier, A. and Rosenberg, N. (1996) Agricultural options for mitigation of greenhouse gas emissions. In: *Climate Change In 1995. Impacts, Adaptations and Mitigation of Climate Change: Scientific-Technical Analyses* (Eds R.T. Watson, M.C. Zinyowera, R.H. Moss and D.J. Dokken). Cambridge Univ. Press, Cambridge, pp. 745–771.

Conway, G.R. and Pretty, J.N. (1991) *Unwelcome Harvest: Agriculture and Pollution*. Earthscan, London.

Culpin, C. (1992) *Farm Machinery*. 12th Edition. Blackwell Scientific Publications, Oxford.

De Klein, C.A.M., McTaggart, I.P., Smith, K.A., Stevens, R.J., Harrison, R. and Laughlin, R.J. (1999) Measurement of nitrous oxide emissions from grassland soil using photo-acoustic infra-red spectroscopy, long-path infra-red spectroscopy, gas chromatography, and continuous flow isotope-ratio mass spectrometry. *Communications in Soil Science and Plant Analysis* **30**, 1463–1477.

De Klein, C.A.M., Sherlock, R.R., Cameron, K.C. and Van der Weerden, T.J. (2001) Nitrous oxide emissions from agricultural soils in New Zealand – a review of current knowledge and directions for future research. *Journal of The Royal Society of New Zealand* (in press).

DETR (1998) *UK Climate Change Programme – Department Of The Environment Transport and the Regions Consultation Paper*. Department of the Environment Transport and the Regions, HMSO, London.

Elsom, D. (1992) *Atmospheric Pollution: A Global Problem*. 2nd Edition. Blackwell, Oxford.

Fowler, D., Sutton, M.A., Skiba, U. and Hargreaves, K.J. (1996) Agricultural Nitrogen and Emissions to the Atmosphere. In: *Issues In Environmental Science and Technology 5: Agricultural Chemicals and the Environment* (Eds R.E. Hester and R.M. Harrison). The Royal Society of Chemistry, Letchworth, pp. 57–84.

Grace, P.R., Oades, J.M., Keith, H. and Hancock, T.W. (1995) Trends in wheat yields and soil organic carbon in the permanent rotation trial at the Waite Agricultural Research Institute, South Australia. *Australian Journal of Experimental Agriculture* **35**, 857–864.

Graedel, T.E. and Crutzen, P.J. (1989) The changing atmosphere. *Scientific American* (September), 28–36.

Grant, R.F., Juma, N.G., Robertson, J.A., Izaurralde, R.C. and McGill, W.B. (2001) Long-term changes in soil carbon under different fertilizer, manure and rotation: testing the mathematical model *ecosys* with data from the Breton plots. *Soil Science Society of America Journal* **56**, 205–214.

Hill, R.A. (2000) *Emission, Dispersion and Local Deposition of Ammonia Volatilised From Farm Buildings and Following the Application of Cattle Slurry to Grassland*. Unpublished Thesis, University of Plymouth.

INDITE (1994) *Impacts of Nitrogen Deposition In Terrestrial Ecosystems*. Department of the Environment, London.

Jarvis, S.C. (1993) Nitrogen cycling and losses from dairy farms. *Soil Use and Management* **9**, 99–105.

Jarvis, S.C. and Pain, B.F. (1990) Ammonia volatilisation from agricultural land. *Proceedings of the Fertilizer Society* **298**, 1–35.

Kasimir-Klemedtsson, Å., Klemedtsson, L., Berglund, K., Maritikainen, P., Silvola, J. and Oenema, O. (1997) Greenhouse gas emissions from farmed organic soils: a review. *Soil Use and Management* **13**, 245–250.

Kasting, J.F. (2001) The carbon cycle, climate, and the long-term effects of fossil fuel burning. In: *Consequences. The Nature and Implications of Environmental Change*. URL: http://www.gcrio.org/CONSEQUENCES/vo14no1/carbcycle.html

MAFF (1997a) *Careful slurry spreading can cut ammonia emissions*. Environmental Protection. R&D Newsletter. Ministry of Agriculture, Fisheries and Foods, London, No. 2, P. 6.

MAFF (1997b) *This little piggy had low crude protein*. Environmental Protection R&D Newsletter. Ministry of Agriculture, Fisheries and Foods, London, No. 1, p 8.

MAFF (1998) *Code of Good Agricultural Practice for the Protection of Air*. Ministry of Agriculture, Fisheries and Foods, London.

MAFF (2000a) *Climate Change and Agriculture In The United Kingdom*. Ministry of Agriculture, Fisheries and Foods, London.

MAFF (2000b) *Towards Sustainable Agriculture: Pilot Set of indicators*. Ministry of Agriculture, Fisheries and Food, p. 30.

MAFF (2001a) *Reducing ammonia emissions – is it practical*. Environmental R&D Newsletter. Ministry of Agriculture, Fisheries and Foods, London, No. 6, p. 5.

MAFF (2001b) *Floating success in reducing ammonia*. Environmental R&D Newsletter. Ministry of Agriculture, Fisheries and Foods, London, No. 7, p. 5.

MAFF (2001c) *An inventory of nitrous oxide emissions from UK agriculture*. Environmental R&D Newsletter. Ministry of Agriculture, Fisheries and Foods, London, No. 7, p. 6.

Minami, K., Horiuchi, H. and Tsubota, K. (1998) Global warming and sustainable agriculture. The 4th JIRCAS International Symposium. *Sustainable Agriculture Development Compatible with Environmental Conservation In Asia*. JIRCAS-International-Symposium-Series, No. 6, pp. 83–98.

Misselbrook, T.H., Van Der Weerden, T.J., Pain, B.F., Jarvis, S.C., Chambers, B.J., Smith, K.A., Phillips, V.R. and Demmers, T.G.M. (2000) Ammonia emissions factors for UK agriculture. *Atmospheric Environment* **34**, 871–880.

Mosier, A.R., Duxbury, J.M., Freney, J.R., Heinemeyer, O. and Minami, K. (1998) Assessing and mitigating N_2O emissions from agricultural soils. *Climatic Change* **40**, 7–38.

Neue, H.U. (1997) Fluxes of methane from rice fields and potential for mitigation. *Soil Use and Management* **13**, 258–267.

Oenema, O., Velthof, G.L., Yamulki, S. and Jarvis, S.C. (1997) Nitrous oxide emissions from grazed grassland. *Soil Use and Management* **13**, 288–295.

Pain, B.F., Misselbrook, T.H. and Chadwick, D.R. (1999) Controlling losses of nitrogen as ammonia from manures. In: *Accounting for Nutrients, A Challenge for Grassland Farmers in ihe 21st Century* (Ed. A.J. Corrall). BGS Occasional Symposium No. 33. British Grassland Symposium, Oxford, pp. 93 98.

Paustian, K., Andrén, O., Janzen, H.H., Lal, R., Smith, P., Tian, G., Tiessen, H. and Van Noordwijk (1997) Agricultural soils as a sink to mitigate CO_2 emissions. *Soil Use and Management* **13**, 230–244.

Pearson, J. and Stewart, G.R. (1993) The deposition of atmospheric ammonia and its effects on plants. *New Phytologist* **125**, 283–305.

RCEP (1996) *Sustainable Use Of Soil.* Nineteenth Report of the Royal Commission on Environmental Pollution (Cm 3165). HMSO, London.

Robertson, G.P., Paul, E.A. and Harwood, R.R. (2000) Greenhouse gases in intensive agriculture: contributions of individual gases to the radiative forcing of the atmosphere. *Science* **289**, 1922–1925.

Robinson, R. (1999) The environment protection viewpoint. In: *Accounting for Nutrients, A Challenge for Grassland Farmers in the 21st Century* (Ed. A.J. Corrall). BGS Occasional Symposium No. 33. British Grassland Symposium, Oxford, pp. 127–136.

Rowell, D.L. (1994) *Soil Science, Methods and Applications.* Longman Scientific and Technical, Harlow.

Rund, T. (1994) Cleaner air putting more crops at risk from sulphur deficiency. *Arable Farming* (December), 22–25.

Schlesinger, W.H. (1997) *Biogeochemistry, an Analysis of Global Change.* 2nd Edition. Academic Press, San Diego, pp. 393–396.

Scott, A., Ball, B.C., Crighton, L.J. and Aitken, M.N. (2000) Nitrous oxide and carbon dioxide emissions from grasslands amended with sewage sludge. *Soil Use and Management* **16**, 36–41.

Simojoki, A. and Jaakkola, A. (2000) Effect of nitrogen fertilization, cropping and irrigation on soil air composition and nitrous oxide emission in a loamy clay. *European Journal of Soil Science* **51**, 413–424.

Smith, K.A. (1997) Soils and the greenhouse effect. *Soil Use and Management* **13**, 229.

Smith, K.A., McTaggart, I.P. and Tsuruta, H. (1997) Emissions of N_2O and NO associated with nitrogen fertilization in intensive agriculture, and the potential for mitigation. *Soil Use and Management* **13**, 296–304.

Smith, P., Powlson, D.S., Smith, J.U., Falloon, P. and Coleman, K. (2000) Meeting the UK's climate change commitments: options for carbon mitigation on agricultural land. *Soil Use and Management* **16**, 1–11.

Sommer, S.G. and Moller, H.B. (2000) Emission of greenhouse gases during composting of deep litter from pig production-effect of straw content. *Journal Agricultural Science* **134**, 327–335.

Speirs, R.B. and Frost, C.A. (1987) The enhanced acidification of a field soil by very low concentrations of atmospheric ammonia. *Research and Development in Agriculture* **4**, 83–86.

Sutton, A.L., Kephart, K.B., Verstegen, M.W.A., Canh, T.T. and Hobbs, P.J. (1999) Potential for reduction of odorous compounds in swine manure through diet modification. *Journal of Animal Science* **77**, 430–439.

Tisdale, S.L, Nelson, W.L., Beaton, J.D. and Halvin, J.L. (1993) *Soil Fertility and Fertilizers.* 5th Edition. Prentice Hall, New Jersey.

Van der Erden, L, De Vries, W. and Van Dobben (1998) Effects of ammonia deposition on forests in the Netherlands. *Atmospheric Environment* **32**, 525–532.

Velthof, G.L. and Oenema, O. (1995) Nitrous oxide fluxes from grassland in the Netherlands: II. Effects of soil type, nitrogen fertilizer application and grazing. *European of Journal of Soil Science* **46**, 541–549.

White, R.E. (1997) *Principles and Practice of Soil Science, The Soil as a Natural Resource.* 3rd Edition. Blackwell Science Ltd, Oxford.

Young, A. (2000) *Environmental Change in Australia Since 1788.* 2nd Edition, Oxford University Press, Oxford.

Pesticides 7

7.1 INTRODUCTION

Pesticides have become an integral part of modern farming, with most crops receiving at least one and usually many more applications. In the UK, cereals are sprayed on average over 6 times during the year (CPA, 2000a). As the value of a crop increases, so does the amount of pesticide used with 10–15 applications per year being normal for some vegetables and fruit. Many of these spray applications include more than one 'active ingredient'. Each active ingredient is, by design, a biocide whose value lies in its ability to kill organisms (specifically those pests, diseases and weeds that are noxious or unwanted in agricultural production).

Pests and diseases cause a 20–40% loss of world-wide crop production. These losses may occur at all stages of the food chain; during harvesting, drying, storage, processing and retailing. The average world grain loss is estimated to be in the region of 20%, although in the UK it is much lower at 10% (Mazaud, 1997). Pesticides reduce attack by pests, diseases and weeds and contribute to higher yields, increased quality and higher economic returns. The use of pesticides is one of the most important contributors to increased agricultural production since the 1940s.

Pesticide history

Pesticide usage dates back to Roman and Greek times, but it was not until the mid- to late-nineteenth century that their use became widespread. Up to the Second World War, the chemical control of pests and diseases involved relatively few substances, confined mainly to high value produce such as fruit, hops and glasshouse vegetables (RCEP, 1979). These early pesticides were mainly inorganic compounds, including elemental sulphur used as a fungicide and simple salts of arsenic, copper, mercury and iron used as fungicides and insecticides. Organic compounds included industrial by-products, like tar distillates, and insecticidal plant extracts such as derris, nicotine and pyrethrum. Few of these pesticides were targeted at particular pests and most were highly toxic and dangerous to use.

It was during the late 1930s and 1940s that the number and complexity of chemicals developed for crop protection began to increase. The insecticides DDT and HCH, and the hormone-type herbicides 2,4-D and MCPA, were first introduced in the late 1940s, followed by the insecticides dieldrin and aldrin in the 1950s (Hassall, 1990). Some pesticides developed at this time were not to be used commercially for another 30–40 years; the first systemic organophosphate insecticide, known as 'Schraden', was produced by German scientists in 1941, but was not widely used because of its toxicity to mammals (Cremlyn, 1979). Instead, the development of organophosphates was pursued principally for military purposes during the 1940–1950s and they did not emerge as the leading type of systemic insecticide for another 30 years.

Pesticide use

Pesticide production is an enormous global industry with a market worth well over £31,000 million (CPA, 2000a). Global pesticide usage in 1999 is summarised in Table 7.1, showing that formulations that control weeds and insecticides are the types most commonly used. Over 60% of all pesticides are applied to staple food crops such as vegetables, cereals, rice and maize (CPA, 2000a). The largest proportion of global usage (55%) is concentrated in the intensive food production systems of North America and Western Europe, although significant increases in sales are being observed in Eastern Asia, notably Thailand, Indonesia and India.

Table 7.1 Breakdown of global agrochemical usage and market value in 1999 by geographical region and product type (CPA, 2000a).

Geographical region	Percent value	Product type	Percent usage
North America	29.4	Herbicides	49.6
East Asia	25.3	Insecticides	26.2
West Europe	22.4	Fungicides	19.5
Latin America	15.3	Others	4.7
East Europe	2.9		
Rest of the World	4.7		

In Western Europe, over 80 million hectares of land are cultivated. Cereals account for 50% of this area, almost all receiving herbicide, 60–80% as fungicide and 15–98% as insecticide (Bigler *et al.,* 1992). Each hectare of cereal crop in the UK receives 3.8 kg of pesticide, usually being made up of 10 different active ingredients (Boatman *et al.,* 1999). Some of the most intensive pesticide used in the world was found in the Netherlands with an average 20 kg of active ingredient applied per hectare of arable land in 1987 (Hurst *et al.,* 1992). This has fallen by 5% since 1991 due to an active policy of pesticide reduction (Eurostat, 1998). Annual usage of pesticides varies considerably and is influenced to a large extent by weather conditions that influence the spread of plant diseases, fungi and insects (CPA, 2000a).

Domestic and export sales of pesticides by UK manufacturers in 1999 totalled £1595 million (CPA, 2000a) compared to approximately £150 million in the late 1970s. The actual rates of pesticide application to UK farmland are gradually declining although the spray area, now exceeding 40 million hectares, is increasing (MAFF, 2000d). Pesticide usage in the UK (discounting sulphuric acid) has declined by over 19% in the last 10 years. This is attributed to moves to newer, more 'active' molecules applied at lower rates coupled with adoption of reduced rate application, particularly of fungicides where disease pressure is low (MAFF, 2000d). The trend in the decline in usage looks to continue with the stated UK government policy of limiting pesticide use '...to the minimum necessary for the effective control of pests compatible with the protection of human health and the environment' (DoE/MAFF, 1995).

This chapter describes the different types of pesticides and their modes of action, identifies the pollution problems that they may cause, and discusses practical options for minimising their environmental impact.

7.2 PESTICIDES AND THEIR APPLICATION

Pesticides are used to control a wide range of agricultural pests, diseases and weeds (Table 7.2). They are also present in a number of veterinary products to control some of the ecto- and endo-parasites encountered in livestock farming. Pesticide products contain a number of constituents, including the active ingredient that kills or controls the target organism as well as a number of additives. These additives include solvents, surfactants, liquid or solid carriers, safeners (to reduce the potential of a pesticide harming the crop itself), and adjuvants (added to a pesticide to increase its efficiency).

Table 7.2 Common pesticides and their target organisms.

Type of pesticide	Target organisms	Examples of active ingredients
Bactericide	Bacteria	Metiram, Difolatan
Fungicide	Fungi	Phenylamides, Dicarboxamides
Herbicide	Plants	Atrazine, 2,4-D, Acifluorfen
Insecticide	Insects	Dimethoate, Permethrin
Molluscicide	Snails and slugs	Methiocarb, Carbaryl
Nematicide	Nematodes	Aldicarb, Dibromochloropropane
Veterinary	Animal parasites	Fenthion, Flumethrin

Pesticides are usually sold in either a liquid or solid formulation, including:

Aqueous Concentrate A concentrated solution of the active ingredient, or ingredients, in water.

Emulsifiable Concentrate A homogenous liquid formulation that forms an emulsion on mixing with water.

Suspension Concentrate A stable suspension of finely ground active ingredient in water intended for dilution before use.

Water Soluble Powder A powder formulation which forms a true solution of the active ingredient, or ingredients, when dissolved in water.

Wettable Powder A powder formulation that is dispersible in water to form a suspension.

Water Dispersible Powder Similar to wettable powder, but involving a more advanced formulation.

Granules Granules or pellets which contain, or are coated with, the active ingredient or ingredients.

Dusts A fine powder formulation used for specific applications.

Active ingredients may be known by their chemical name, but usually have a more straightforward 'common' name. A single active ingredient can also occur in many different brand-named products. For example, tetramethylthiuram disulphide is a fungicide used for controlling damping off in seedlings, is commonly known as thiram, and is marketed by several companies in the UK in different brand-named products.

Pesticides are usually applied by spray (as a solution, emulsion or particles), spread (as granules or dust) or dispersed as a fumigant (Hassall, 1990; Culpin, 1992). Herbicides are generally (though not exclusively) applied once during the growing season with timing of application depending upon the weed problem, whilst fungicides and insecticides are more likely to be applied several times during the crop production cycle (including seed treatment and post-harvest storage). Pesticides are usually applied by dilution with water and application as a fine spray with a tractor-mounted or trailed sprayer. The spray is usually applied downwards directly onto the crop, except in orchards where the spray is directed upwards under high pressure to reach fruit at the top of the trees. Aerial crop spraying plays an important role in many countries (notably the US and Australia), but is relatively limited in the UK, accounting for only 0.02% of the total treated area (MAFF, 1998a). Solid formulations of fungicide and insecticide may also be applied both at the start (as a seed dressing) and end (as a dusting on stored grain) of the production cycle and grain stores may be fumigated with insecticide prior to harvest. Granular herbicides, fungicides and insecticides are also available for direct-to-soil application.

7.3 MODE OF ACTION

The way in which a pesticide acts varies and is often highly complex, but toxicity usually takes effect by interfering with the biochemical processes of the pest or 'target' organism (Baird, 1995). Chemical substances may produce one of two types of toxic effect:

- acute effects that have a rapid onset over a short period of time with pronounced symptoms;
- chronic effects that persist over a long time period, generally occurring when an organism is exposed to repeated sub-lethal doses.

A widely used measure of comparative acute toxicity is the LD_{50}. This is the median lethal dose of a given substance at which 50% of the subjects (commonly male rats) die (Wentz, 1989). Some examples of LD_{50} values are given in Tables 7.3 to 7.5, and are expressed as mg of substance per kg body weight of the subject. The lower the LD_{50} value, the more toxic the substance is to the test organism. This information may be adapted to give an indication of toxicity to humans or livestock (Hassall, 1990). The toxic effect produced by a pesticide is largely determined by its biological activity and the magnitude of dose received by the target organism. When using pesticides, the biologically active ingredient is therefore formulated, prepared and applied to ensure that its availability to the target organism is maximised (Graham-Bryce, 1987).

The modes of action of most pesticides fall into the following basic categories (Hassall, 1990; van Emden, 1992):

Poison Ingested by the pest organism before releasing toxins into its stomach.

Contact Applied directly to the pest organism, penetrating its surface and producing a localised toxic effect. Usually they remain active for a few days at most.

Residual Act in the same way as contact pesticides, but do not need to be applied directly to the pest organism since they remain active for long periods.

Translocated Active ingredient is mobilised within the pest organism and has a more effective toxic effect upon it.

Systemic Active ingredient is mobilised within the crop or animal being protected and is then transferred to the target pest.

Most pesticides used in agriculture today are synthetic organic chemicals that act by interfering with a vital metabolic process in the organisms to which they are targeted. Commonly affected metabolic processes are photosynthesis (herbicides), energy release and transfer (herbicides, fungicides), nerve cell impulses (insecticides), cell division and growth (herbicides, fungicides) and biosynthesis (fungicides). We now look at the three main classes of pesticides, namely insecticides, herbicides and fungicides in more detail.

7.4 INSECTICIDES

There are hundreds of different compounds commercially available as insecticides, but these fall into three general chemical groups (organophosphates, carbamates and synthetic pyrethroids) which account for over 75% of the market (Conway and Pretty, 1991). Most insecticides work by damaging the nervous system of the target insect, either by affecting neurons or by interfering with the transfer of nerve impulses between neurons.

The first synthetic insecticides to be widely used in agriculture were the organochlorines, the best known being DDT. They were heralded as a major breakthrough and revolutionised pest control, particularly for vectors of diseases such as malaria. The compounds involved were cheap, effective, non-toxic to humans and their persistence and long-lasting effectiveness were initially considered as beneficial. It is now evident that their broad-spectrum insecticidal activity, persistence in the environment and tendency to bioaccumulate along food chains caused significant environmental problems and as a result their use has been drastically reduced. Organochlorine pesticides are generally banned in most western countries although they are still used in developing countries, particularly in the tropics.

One organochlorine compound, lindane, remains of concern in the UK, despite being banned or severely restricted in 37 countries, and being phased out throughout the EU. Lindane (or gamma-HCH as it is also known) is currently approved for use in the UK for non-agricultural uses, such as in wood preservation. Exposure to lindane has been linked with a number of human health issues, including blood disorders, multiple congenital abnormalities and breast cancer.

Organophosphates (OPs)

Organophosphate insecticides are widely used in agriculture and have largely replaced the persistent organochlorine insecticides (Baird, 1995). Organophosphate affect insects by inhibiting the enzyme acetylcholinesterase, which is responsible for hydrolysing acetylcholine, a substance involved in the transmission of nerve impulses.

In common with other animals, insects have a nervous system to transmit impulses around the body. This system is made up of a series of neurons that interconnect via gaps called synapses to form a complex neural network. Nerve impulses are conducted through this network by both electrical (along the neurons) and chemical (across the synapses) processes. The chemical process at a synapse involves acetylcholine. This is produced by each neuron in turn as it is electrically stimulated, diffusing across its adjacent synapse to induce an electrical impulse in the next neuron. In order to avoid hyper-activity, levels of acetylcholine must be precisely controlled by the insect (and indeed all other animals). This is achieved by the enzyme acetylcholinesterase, which hydrolyses acetylcholine as soon as it has completed its task of transferring an impulse across a synapse.

Organophosphate insecticides (and carbamates) work by inhibiting the activity of acetylcholinesterase that leads to an accumulation of acetylcholine within the neural network. As levels of acetylcholine increase the insect becomes continually and increasingly stimulated by nerve impulses leading to hyperactivity, disruption of its behaviour and eventual death.

One of the potential problems with OPs is that although they generally break down in the environment relatively quickly, they are extremely toxic to mammals whilst they remain chemically active. In particular, increasing concern is being voiced about the adverse effects of OP pesticides upon human health.

Table 7.3 General properties of some common insecticides.

	Organophosphates		Carbamates	Pyrethroids	
	Parathion	Malathion	Aldicarb	Permethrin	Deltamethrin
LD_{50} ($mg\,kg^{-1}$)	3.6–13	1375–2800	1	>500	50–500
Log K_{ow}	3.9	2.7	1.6	6.6	
Application	Spray/ Fumigant	Spray/ fumigant	Spray	Various	Various
Solubility ($g\,l^{-1}$)	0.00654	0.145	1–0.1	Insoluble	Insoluble

Carbamates

These insecticides are commonly used to control insects that do not readily respond to the less expensive organophosphates (whiteflies, mealy bugs, cockroaches etc.). Carbamates tend to have a relatively broad-spectrum of activity and also function by inhibiting the enzyme acetylcholinesterase. Like organophosphates, carbamates show little tendency for environmental persistence, but may be acutely toxic to mammals (Table 7.3).

Pyrethroids

These may be either natural or synthetic. Natural pyrethrum is extracted from the flowers of *Chrysanthemum cinerariaefolium* and is still in current use. Due to the inconsistent supply of the flowers the cost of natural pyrethroids tends to be high. This stimulated a search for synthetic pyrethroids and they were first formulated at Rothamsted Experimental Station in the 1970s. Pyrethroids are characterised by having a rapid 'knock down' effect on most insects and act by attacking the central and peripheral nervous system. Pyrethroids have very low mammalian toxicity and limited environmental persistence, due to the fact that the active ingredients (esters of asymmetric acids and alcohols) are rapidly hydrolysed in water and degraded by light.

7.5 HERBICIDES

More is spent globally on herbicides than any other type of pesticide (Table 7.1). Some herbicides are non-selective (e.g. glyphosate) and are used specifically to clear the soil of vegetation. Of the herbicides, the three most widely used groups are the triazines, phenoxyacetic acids and carbamates (Table 7.4). Herbicides are applied directly to soil, just prior, or sometimes during crop growth in order to destroy weeds. Left unchecked, weeds compete with the crop for light, moisture and nutrients thereby reducing growth and yield. Most herbicides need to be selective; their application should only kill the weeds competing with a crop and they should not affect the growth of the crop itself (Hassall, 1990). Selectivity can be achieved in a number of ways, including:

- the herbicides chemical activity;
- its rate of application;
- differences and variations in growth stage between the crop and weed;
- timing of the application.

Table 7.4 General properties of some common herbicides.

	Triazines (Atrazine)	Carbamates (Barban)	Phenoxyacetic Acids (2,4-D)
LD_{50} ($mg\,kg^{-1}$)	1870–3080	600	50–100
Application	Pre-emergent spray	post-emergent spray	Post-emergent spray
Solubility ($g\,l^{-1}$)	0.07	11	0.890

Triazines

This group of herbicides includes atrazine, simazine and cyanazine. Triazines are applied as pre-emergent sprays directly to the soil and function by blocking one of the main biochemical transport systems in the operation of photosynthesis and reducing the effectiveness of the chloroplastic membranes. Typical symptoms of triazine use are chlorosis and desiccation of leaves of the emerging weed seedlings. Atrazine is one of the few herbicides that may pose a potential health risk to mammals, but only when it is transformed in the stomach to a more carcinogenic form (Conway and Pretty, 1991).

Phenoxyacetic acids

Unlike the triazines, phenoxy acid herbicides are applied as post-emergent sprays directly to the growing crop. They are used for the control of dicotyledonous (broad leaf) weeds and function by being absorbed by lipids in the leaf and disturbing the normal growth pattern of the plant by interfering with its genetic make up. The application of 2,4-D causes deformation of growing shoots and leaves and epinasty (twisted, spindly growth) of the stems (Hassall, 1990).

Phenoxy acid herbicides, such as 2,4-D degrade rapidly on contact with soil and generally show limited toxicity to humans. Some commercial formulations of phenoxy acids have, however, been susceptible to contamination with dioxin in the past. In the 1970s considerable amounts of dioxin were found in 2,4,5-T (no longer used in the UK) and are known to have caused birth defects, skin disease and neurological disorders in human populations exposed to the herbicide (Conway and Pretty, 1991).

Carbamates

This group of herbicides is often used to control grassy weeds in cereal crops. Barban, a commonly used carbamate herbicide, both controls wild oats (*Avena fatua* and *A. ludoviciana*) and acts as a growth retardant. It has limited persistence of 2–4 weeks (being broken down by micro-flora present in the soil) and the timing of application is vitally important to ensure weed removal. The action of this herbicide is thought to be via interference with protein synthesis and formation (Hassall, 1990).

7.6 FUNGICIDES

The use of fungicides to control crop diseases involves a different approach from that used for the control of insects and weeds. There are three major types of fungicide currently in use: traditional inorganics, phenylamides and carboxamides (Table 7.5). Relatively little is known about their detailed mode of action.

Table 7.5 General properties of some common fungicides.

	Traditional bordeaux mixture	Phenylamides metalaxyl	Carboxamides carboxin
LD_{50} (mg kg^{-1})	700	–	3200
Application	Foliar spray	foliar spray soil drench root/soil dressing	seed dressing
Solubility (g l^{-1})	Insoluble	8.4	101.0

Fungi have a considerable ability to regenerate and spread rapidly from a relatively small population base. The application of fungicide to a crop already well infected with a disease is therefore futile, as considerable damage to the crop will already have occurred. The most effective approach to disease control is to treat the crop either before the arrival of the infection, or in the early stages of its development. This may be done by the use of contact or systemic fungicides applied as soil drenches, seed and root dressings or foliar sprays.

Traditional inorganics

Inorganic fungicides, notably transition metal complexes such as copper based Bordeaux mixture, mercurous chloride and triphenyltin have long been used to control fungal diseases on crops like grape vines, fruit trees, hops, potatoes, brassicas and onions. Bordeaux mixture is a combination of copper sulphate ($CuSO_4$) and slaked lime ($Ca(OH)_2$) forming a gelatinous precipitate that can be used against a relatively wide variety of fungal diseases, either in powder/dust form, or mixed with water and applied through a sprayer. On contact with fungi, the copper in the precipitate is solubilised and complexed by exudates released by the fungal spores. These complexes are then transported to the fungal spores where they exert a toxic effect.

Phenylamides

These fungicides include metalaxyl, ofurace and oxadixyl. They are used extensively in the control of a certain group of fungi known as oomycetes, which are systemic within the host plant. Oomycetes produce motile spores that require water for their transfer to the host organism at the infective stage. Metalaxyl is a fungicide that has a specific action against oomycetes, particularly late blight in potatoes (*Phytophthora infestans*) and downy mildews. It functions by interfering with enzymes that control regeneration and spore formation, initially stopping growth, then killing the organism.

Carboxamides or oxathins

These are used extensively against rusts and smuts in cereal crops. Like many other fungicides and herbicides, carboxamides display relatively limited mammalian toxicity. Carboxin, one of the most common types of carboxamides, inhibits enzymes involved in the metabolism and respiration of fungi such as loose smut (*Utstilago nuda*).

7.7 THE CAUSES OF PESTICIDE POLLUTION

Pesticides as pollutants

The environmental impact of different pesticides is linked to the method, form and timing of their application (Miller, 1991) and their effect on the natural environment varies greatly. The environmental consequences of pesticide use first became the object of widespread public interest in the 1960s, notably through the publication of Rachel Carson's famous book, Silent Spring (Carson, 1985; first published in 1962). The use of pesticides in agricultural systems has expanded primarily due to:

- relatively low cost and high returns, commonly combined with external financial incentives in the form of subsidy;
- the intensification of agriculture, including the introduction of high yielding varieties requiring greater pesticide application;
- bad practice due to a lack of understanding regarding the consequences of pesticide use;
- a tendency to try and achieve total protection from pests ('cosmetic control') in high value crops, such as fruit and vegetables.

An 'ideal' pesticide would only affect its target organism, be non-persistent and have no harmful environmental effects (Conway and Pretty, 1991). However, most pesticides do not have these traits and consequently they may be directly toxic to non-target organisms, accumulate in ecosystems or cause ecosystem disruption. Pesticides may cause point source or diffuse pollution via the following pathways:

- by direct contamination of ground and surface waters;
- by soil contamination and subsequent leaching into ground and surface waters, surface run-off and/or soil erosion;
- by direct contamination of non-target organisms;
- by contamination of non-target organisms by residues or persistent compounds.

Point source pollution is often caused by poor agricultural practice and includes leaking storage facilities, spillages whilst filling equipment, run-off from wash-down areas during cleaning, and the inappropriate disposal of unused pesticides (Miller, 1991; MAFF, 1999). Diffuse pollution may occur even when good agricultural practices are adopted because a significant proportion of the pesticide applied inevitably does not reach its target organism (Helweg, 1994; Brown *et al.*, 1995). In a review of published work on arthropod pests, crop pathogens and weeds, Pimentel (1995) concluded that less than 0.1% of pesticides applied reach their target organism.

Resistance and resurgence

When regularly and widely exposed to a particular pesticide, pest organisms often build-up resistance by a process of selection. Insects may develop resistance by developing less permeable cuticles, retaining toxins in fatty tissue more effectively or by better enzyme systems metabolising the toxin (van Emden, 1992). Resistant individuals are usually rare in a pest population, and so initially pesticides are often very effective. Continued use, however, favours the resistant individuals and the efficacy of the pesticide may be reduced (Figure 7.1). Although the short life cycles and large numbers of insects make them particularly adept at developing resistance, it is also evident in many other organisms including plants, mammals and fungi (Box 7.1).

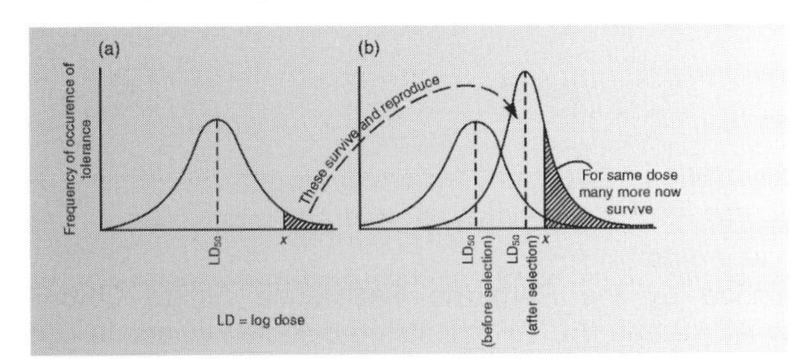

Figure 7.1 Development of resistance. (a) Distribution of tolerances before selection; (b) Distribution of tolerances after selection compared with distribution before selection (reproduced with kind permission van Emden, *Pest Control*, 1992. Cambridge University Press).

As individual pesticides become ineffective, many farmers are inevitably forced onto a 'pesticide treadmill'; applying more and more pesticides to try and maintain yields as the resistance of the target pest increases.

Another consequence of pesticide usage is the resurgence of pest populations, particularly those with resistance. This phenomenon is caused by application of broad-spectrum pesticides that control pests, but which also destroy their natural enemies at the same time. In the absence of natural enemies, populations of pests can increase rapidly, causing a bigger problem than would occur if the pesticide had not actually been applied.

Box 7.1 Pesticide resistance

Fungicides – powdery mildew and Botrytis cinerea

Svec *et al.* (1995) detected resistance of Wheat Powdery Mildew (*Erysiphe graminis*) to the fungicides triadimenol, tebuconazole, propiconazole, flutriafol and fenpropimorph sampled from the Czech Republic, Austria, Hungary and Slovakia. Resistance to triadimenol was the highest, having an MRF (mean resistance factor) 29 times higher than non-resistant standards. Stehmann and de Waard (1995) showed that the fungal plant pathogen *Botrytis cinerea* has developed 100% resistance to the commonly used benzimidazoles and dicarboximides fungicides.

Insecticides – the housefly

The housefly, *Musca domestica*, is a serious pest in intensive livestock units and has shown increased resistance to a range of commonly used insecticides. Pap and Farkas (1994) conducted a study to assess the resistance of different populations of houseflies to a range of insecticides. Resistance was identified by calculating resistance ratios, a relative measure comparing the effect of insecticides. The percentage of populations exhibiting evidence of resistance were:

Active Ingredient	Percentage
DDT	60
methoxychlor	50
lindane	13
malathion	83
permethrin	63
deltamethrin	79
diflubenzuron	0

Herbicides – wild oats

A survey funded by the Herbicide Resistance Action Committee and conducted by ADAS and IACR-Rothamsted detected evidence of resistance in winter wild oats (*Avena ludoviciana*) on farms in Essex (Rush, 1994). Samples were tested with a range of herbicides under controlled conditions and resistance was found to a number of herbicides commonly used to control the weed, including fenoxaprop-ethyl, diclofop-methyl and fluazifop-P-butyl.

7.8 BEHAVIOUR AND FATE OF PESTICIDES IN THE ENVIRONMENT

Before discussing the key problems associated with the use of pesticides in modern agriculture, it is useful to consider those factors that determine their behaviour in the environment. These factors have an important influence upon the fate and impact of pesticides as pollutants.

Persistence

After application, a pesticide is subject to varying degrees of biological and chemical breakdown (Alloway and Ayres, 1993; Harrad, 1996) and its persistence depends upon its ability to resist this degradation. Generally, naturally derived pesticides, such as pyrethrum have relatively low environmental persistence, breaking down rapidly when exposed to sunlight (photo-instability). In contrast, many synthetic pesticides can be highly persistent (Warner, 1995). Persistence can be defined as a measure of the time necessary to degrade the pesticide into harmless products (Young *et al.,* 1992) and does not necessarily imply that environmental problems will occur. Some pesticides may persist for long periods without causing environmental harm as a pollution pathway to a sensitive site or target may not exist.

Other pesticides may breakdown very rapidly, but create intermediate metabolites that present a greater environmental hazard than the original compound. Some of the most notorious metabolites are DDD and DDE from the chemical/microbial dechlorination and photochemical reactions of DDT. In particular, DDE is thought to be a strong androgen receptor (with the potential to cause male sterility), with persistence exceeding that of DDT (Aislabie *et al.,* 1997). Further examples of potentially hazardous metabolites include 1,2,4-triazol, a metabolite of several fungicides, and chlorallyl alcohol, a metabolite of the nematicide 1,3-dichloropropene which have been found in drinking water in the Netherlands (Loch, 1991).

Useful indicators for the behaviour of pesticides in the environment are aqueous solubility and the partition coefficient (Tables 7.3 to 7.5). The partition coefficient (K_{ow}), usually expressed as a log value, is a measure of the distribution of the pesticide between liquid layers of water and an alcohol, 1-octanol:

$$K_{ow} = \frac{\text{concentration of pesticide in octanol}}{\text{concentration of pesticide in water}}$$

Both the solubility in water and partition coefficient of pesticides vary considerably and are believed to have a considerable bearing on environmental persistence (Munz and Bachmann, 1993). For example, DDT has a solubility in water of only 0.0034 $\mu g\ ml^{-1}$ and a log K_{ow} of 3.9–6.2. Consequently, its residence time in soil, in some situations is estimated to be greater than 20 years (Van Zwieten *et al.,* 2001). In contrast another organochlorine, aldicarb, has a solubility of over 6000 $\mu g\ ml^{-1}$, a log K_{ow} of 1.6 and a residence time in soil of only several weeks.

As the value of log K_{ow} increases a pesticide tends to become more hydrophobic and it is more likely to bind to organic molecules. Pesticides with partition coefficient values of greater than 7 do not generally move far in the agroecosystem. Instead, they tend to bind strongly to the organic matter in soils and sediments. However, log K_{ow} values of between 4 and 7 indicate that the chemical is lipophilic in nature and consequently has the tendency to dissolve and accumulate in fatty tissues. If bioaccumulation occurs then this effect may be amplified through food chains, putting organisms at higher trophic levels (usually carnivores) at risk (Baird, 1995; Turnbull, 1996).

In some cases, the inherent lipophilicity of pesticides means that they may be adsorbed to plastics, rubber and certain polythenes. Such a phenomena was blamed for the delay in cleaning a plastic drinking water supply system after an incident in Denmark, where it took over one month of flushing to reduce concentrations of the OP, parathion, to safe levels (Helweg, 1994).

Some of the inorganic compounds traditionally used as pesticides are particularly recalcitrant and may persist in the environment for long periods (Young *et al.,* 1992). Although they do not bioaccumulate, repeated use can lead to accumulation in soil. An example of this is the traditional copper-based fungicide Bordeaux Mixture. Since inorganic compounds are usually applied in the very early stages of infection they require a reasonable degree of persistence in order to be successful. Copper-based fungicides have extraordinarily long persistence times and residues from treatment may be present in soils for over 100 years (Merry *et al.,* 1986). There is concern, therefore, that repeated applications can lead to copper accumulation in the soil, crop toxicity, a decline in beneficial soil organisms, and eventual yield loss (Filser *et al.,* 1995; Helling *et al.,* 2000).

The fate of pesticides in soil

Pesticides may enter soil profiles by being applied directly as drenches, granules or sprays, as run-off from plant canopies, or via seed dressings. The subsequent fate of these pesticides is then influenced by the interaction of its persistence and degradation with a number of processes, including biochemical degradation, volatilisation, adsorption, leaching and plant uptake.

The major mode of pesticide degradation in soil is biochemical due to the action of soil microbes. The susceptibility of a synthetic pesticide to microbiological breakdown is largely related to its structure, notably its similarity to the micro-organisms usual organic substrate. Large ring structures (e.g. DDT and parathion) and complex long chain organic molecules are broken down less rapidly than linear chain structures. Furthermore, the presence of chlorine substituents on some molecules may increase environmental persistence. For example, DDT is considered to be one of the most persistent pesticides, but its non-chlorinated analogue diphenylmethane, is readily biodegradable (Aislabie *et al.,* 1997). In some cases exposure to pesticides over a relatively long time period may induce certain soil bacteria to develop the ability to break down some of the more complex chemical structures (Herrling *et al.,* 1993; Megharaj *et al.,* 2000). This can lead to a reduction in their efficacy to such an extent that alternative pesticides or other control practices are required (RCEP, 1996). Once a pesticide has penetrated below the biologically-active upper soil horizons and into the underlying parent material its rate of degradation declines significantly. Pesticide persistence in subsoils may be between 5 and 20 times greater than in surface horizons (Helweg, 1994).

Another factor that influences the activity and biodegradability of pesticides is their affinity for adsorption by soil organic matter. Soil organic matter possesses an array of chemical functional groups (e.g. hydroxyls, carboxyls, phenolics and amines) which can interact with pesticides (Young *et al.,* 1992). As many pesticide molecules are non-ionic, non-polar and generally hydrophobic, organic matter provides important sites for their adsorption (Harrad, 1996). Adsorption is the process whereby a chemical moves from the liquid or solution phase to the solid phase. The adsorption characteristics of a pesticide for organic matter may be estimated by using log K_{ow}. Adsorption of a pesticide onto organic matter may affect its behaviour and fate in the soil in a number of ways:

- it may make the pesticide physiologically inactive or more susceptible to degradation by microbial action;
- by reduced mobility in the soil making the pesticide less prone to leaching (Pedersen *et al.*, 1995);
- by enhanced mobility of the pesticide. Dissolved organic matter or colloidal particulate matter (colloids are particles of <1 μm in diameter) can form complexes with pesticides, greatly increasing their susceptibility to leaching loss (Gerstler, 1991);
- pesticides associated with organic matter are susceptible to soil erosion and the movement to watercourses as suspended load (Brown *et al.*, 1995). It is thought that soil erosion is responsible for the continued reappearance of many banned pesticides (e.g. aldrin and dieldrin) in surface waters in the UK even though they are no longer applied (Environment Agency, 2000).

The clay content of soils may also greatly influence the fate of agricultural pesticides. Clay-sized particles (<2μm), in particular alumino-silicate minerals, have important properties which explain their fundamental importance in soil chemistry. They characteristically have very large surface areas (700–800 $m^2\ g^{-1}$ for montmorillonite an expanding lattice clay) and can carry a permanent negative electrical charge. This means they are of considerable importance in the adsorption of ionic and ionizable pesticides. Many of the triazine herbicides, for example, are weak bases in acid media and one of the amino groups may become protonated, therefore enhancing its adsorption by clays at low soil pH.

Loss of pesticides from soil

Leaching losses of pesticides from soil typically occur when the prevailing rainfall (or irrigation) exceeds evapotranspiration losses and the soil water content rises above field capacity (White, 1997; Marshall *et al.*, 1999). At this point, drainage water starts to move down the soil profile carrying chemicals, such as pesticides with it. The susceptibility of a pesticide to leaching in this manner depends upon the interaction of the chemical properties of the pesticide (i.e. how easily it is degraded, adsorbed or volatilised) with the chemical and physical properties of the soil.

Any factor that reduces soil microbiological activity (e.g. low organic matter, oxygen, temperature or soil moisture) will decrease the potential rate of pesticide degradation and maintain the availability and susceptibility of pesticides to leaching (Young *et al.*, 1992; Harrad, 1996). An example of how a reduction in soil microbial activity may affect degradation of applied pesticides was observed in regions of Denmark (Helweg, 1994). In the winter months, a mean precipitation surplus of only 320 mm coupled with a mean soil temperature of 3°C slowed the microbial degradation of pesticides significantly and increased the risk of leaching and the pollution of ground and surface waters.

Theoretically, a pesticide applied to a clay soil with a high organic matter content will be less susceptible to leaching than if applied to a sandy soil with a low organic matter content. In practice, however, the movement of pesticides through heavy soils is influenced greatly by the occurrence of 'by-pass flow'. This may be defined as accelerated movement of water and dissolved chemicals down the soil profile through cracks, macro-pores, animal burrows and old root channels in the upper soil horizons (Funari *et al.*, 1991; Marshall *et al.*, 1999). Williams *et al.* (1995) applied seven commonly used pesticides at recommended rates to crops growing on a heavy clay soil and monitored the resultant pesticide concentrations in soil and drainage waters. Although the proportion of the applied pesticide recovered in drainage water was generally very small (Table 7.6), the concentrations of pesticide found during rainfall events were well in excess of the EU limit of 0.1 $\mu g\,l^{-1}$. These researchers concluded that pesticide leaching to streams was considerably more pronounced than expected on the basis of 'classical leaching theory' and that this was due to by-pass flow. Under certain circumstances, therefore, processes such as adsorption and degradation within the soil can be of relatively limited effectiveness in reducing leaching losses (Oliver *et al.*, 2000).

Plant uptake is another pathway that may determine the fate of pesticide residues and its extent is determined by its bioavailability. This is dependent upon many soil and plant factors, as well as the type of pesticide involved (Alloway, 1995; Barber, 1995). Uptake via root systems is achieved if the pesticide has certain water solubility and lipophilicity characteristics (White, 1997). A partition coefficient of log $K_{ow} = 4$ has been observed to be the upper limit for root uptake (Graham-Bryce, 1987).

Table 7.6 Proportions (%) of pesticide applied found in drainage water and maximum concentrations ($\mu g\,l^{-1}$) of pesticide recorded (Williams *et al.*, 1995).

	Maximum pesticide concentration in water ($\mu g l^{-1}$)	Maximum % of applied pesticide found in water
Isoproturon	17.2	0.014
Simazine	15.3	0.092
Dimethoate	3.0	0.00093
MCPA	46.8	0.007
Atrazine	58.4	0.002
Carbofuran	264.0	1.1
Aldicarb	2.8	0.012

A final mechanism for the movement of pesticides from soil (and leaf surfaces) is volatilisation, i.e. the flux of organic chemicals into the ambient air above treated surfaces (Pestemer and Krasel, 1992). The rate and extent of volatilisation depends upon the incident environmental conditions, the method of application, and the volatility of the pesticides involved. The volatilisation of persistent pesticides and their long-range atmospheric re-deposition in precipitation has caused pesticide pollution in many remote lakes and other surface waters (Conway and Pretty, 1991; Siebers *et al.*, 1994) whilst lindane has even been found in arctic air and snow (Cleemann *et al.*, 1995).

7.9 ENVIRONMENTAL PROBLEMS CAUSED BY PESTICIDE RESIDUES

Once pesticides enter the environment they have the potential to cause pollution problems. In this section we consider some of the effects of pesticide contamination.

Pesticides in surface and ground waters

Ground and surface waters are vulnerable to pollution by pesticides and a large number of different active ingredients have been detected during regular surveys of water quality in the UK (MAFF, 2000d; Environment Agency, 2000). Surveys indicate that incidents recorded annually have increased by 35% from the middle to late 90s (Figure 7.2), mainly due to an increase in Category 1 (the most severe) and Category 2 pollution incidents (Environment Agency, 2000).

Acute pollution occurs when pesticides enter surface waters directly due to accidents or bad practice during storage, application or disposal. A common cause of acute pollution are sheep dip chemicals, cypermethrin and diazinon, which are especially toxic to aquatic organisms (Box 7.2). Other common pesticides found in UK surface waters include mecoprop which is an agricultural herbicide (and significantly also used in domestic gardens), and isoproturon, MCPA, 2,4-D, diuron and dichlorprop which are also all relatively environmentally labile herbicides (MAFF, 2000d).

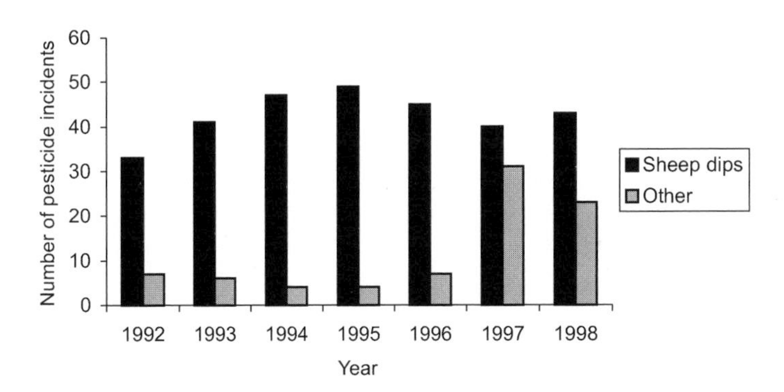

Figure 7.2 Distribution of substantial aquatic pollution incidents from 'other pesticides' and sheep dips between 1992 and 1998 (adapted from Environment Agency, 2000).

Chronic effects are usually from diffuse sources due to pesticides or secondary compounds leaching from soils. Groundwaters are particularly susceptible to chronic pollution and pesticide residues are regularly detected (Table 7.7).

Box 7.2 UK case study of sheep dip pollution

The Environment Agency is responsible for monitoring the water quality of all rivers, lakes and streams in the UK. In addition to routine monitoring, surveys are conducted when pollution incidents occur and this information may be used in prosecutions. The incident described below occurred in the north-west of England during the 1990s.

A pollution inspector was requested to investigate a problem in a small stream in Lancashire. The water quality of the stream had deteriorated due to a spillage of OP sheep dip entering via a surface water drain. A biological survey was conducted to assess the effects of the incident conducted by collecting macro-invertebrates (insects, snails, shrimps, etc.) both upstream and downstream of the point source. The number of invertebrate families living in the stream is an indicator of the water quality of each site. The results of the survey were:

	Presence in sample	
Invertebrate families	Upstream	Downstream
Stoneflies		
Perlodidae	✓	✗
Chloroperlidae	✓	✓
Mayflies		
Heptagenidae	✓	✓
Leptophlebidae	✓	✗
Caddisflies		
Rhyacophilidae	✓	✗
Limnephilidae	✓	✗
Beetles		
Dytiscidae	✓	✗
Elminthidae	✓	✗
Midge larvae		
Chironomidae	✓	✗
Simulidae	✓	✗
Freshwater limpet		
Ancylidae	✓	✓
Worms		
Tubificidae	✓	✓
Lumbriculidae	✓	✗

The results showed that the spillage of sheep dip had a major effect on the animals living in the stream, causing a substantial change in the invertebrate community downstream of the pollution source.

In some cases the risk of pollution may relate to previous, rather than current, pesticide use. Raju *et al.* (1993) studied pesticide residues of atrazine, metolachlor and cyanazine in a clay loam soil following their application at rates of 2.2, 1.2 and 2.5 kg ha^{-1} respectively for a period of 2 to 3 years. Four years after the application of the herbicides, analysis showed that the soils still contained residues and metabolites in sufficient quantities to potentially contaminate surface and ground waters. Metolachlor residues were found to be highest with an average of 103 mg kg^{-1}, while atrazine and cyanazine averaged between 10 and 16 mg kg^{-1}.

Although atrazine was banned in the early 1990s there is little or no indication that concentrations are declining across groundwater monitoring sites, some 10 years later. This indicates the long-term nature of aquifer contamination by pesticides and stresses that the benefits of regulation may take considerable time to occur (Environment Agency, 2000).

Table 7.7 Pesticides detected in ground waters in the UK (of the 130 sampled for) at 8546 locations (Environment Agency, 2000).

Pesticide	Percentage of samples exceeding EU limit of 0.1 $\mu g\,l^{-1}$
Atrazine	11.0
Bentazone	2.1
Isoproturon	2.0
Diuron	1.9
Mecoprop	1.8
Dichlobenil	1.1
Terbutryn	1.0
Chlorotoluron	<1
Trietazine	<1
Imazapyr	<1

The nature of pesticide pollution in the UK varies regionally according to local agricultural production. Most pollution incidents involving crop protection pesticides (e.g. herbicides and fungicides) occur in East Anglia and the Midlands whilst incidents involving sheep dip are mainly to be found in the hill and upland areas of Wales, north-west England and some parts of Scotland.

Effects on non-farmed organisms

Although pesticides are targeted at relatively few specific pest organisms, their application commonly impacts upon the wider agroecosystem, affecting non-target organisms. For example, broad-spectrum insecticides may harm species such as beetles and spiders by direct toxic effect. To reduce broad-spectrum effects, products are being developed which are more specific to their target organisms and increasing care is being taken over pesticide application (CPA, 2000a). Nevertheless, pesticides remain, by design, biologically active chemicals. In the UK, English Nature expressed particular concern about the possible effects on wildlife of long-term exposure to pesticides and concluded that any reduction in pesticide use would benefit nature conservation (NCC, 1990).

Humans are also a category of non-target organism with occupational or accidental poisoning (e.g. pesticide drift) and food residues being possible causes of exposure. We do not consider the human health aspects of pesticides here although in Box 7.3 but provide the reader with evidence that pesticide residues are present within the food chain.

Box 7.3 Pesticide residues

Organochlorine pesticide residues in cheese

Organochlorines can enter the human food chain in a number of ways. One of the more unusual is via grazing livestock; cattle may involuntarily ingest up to 18% of their DM intake as soil. If grazing land has been treated with organochlorines, bioaccumulation may occur. Fat rich products from the livestock, particularly cheese, butter and milk may subsequently contain the pesticide (Wong and Lee, 1997). A study of the accumulation of organochlorine pesticides in commonly consumed cheeses in Spain observed that of the 146 samples collected, 94.5% were observed to contain chlordane although its use had been banned since 1977 (Bentabol and Jodral, 1995). DDT was also observed and occurred in many of the samples with a mean concentration of 55 $\mu g\,kg^{-1}$ and a maximum of 406 $\mu g\,kg^{-1}$. The human health risk was evaluated based on the average per capita consumption of cheese and the results showed that in all cases intakes were well below acceptable daily intakes.

Organophosphorus residues in carrots

OPs, notably chlorfenvinphos, phorate, triazophos, quinalphos and primiphos-methyl, have been increasingly used in the UK over the last 25 years to control carrot fly (Psila rosae). By the mid-1990s, virtually the entire UK carrot crop was routinely sprayed up to a maximum of nine times per year.

At this time, pesticide residues were monitored by a standard procedure involving the analysis of composite (blended) samples of carrots from commercial sources and were generally found to comply with the statutory maximum residue levels (MRLs) set for the UK under the Pesticides (Maximum Residue Levels in Food) Regulations, 1988 (Beaumont, 1992).

In 1995, MAFF announced that researchers reporting to the independent Advisory Committee on Pesticides (ACP) had modified the standard monitoring procedure and analysed carrots individually and found unexpectedly high residues of OP insecticides with 1–2% of carrots tested containing residues up to 25 times higher than the MRL (PSD, 1995). With no explanation for these high levels, the ACP concluded that '...margins of safety have been eroded to a level where action is required to restore them'. As a consequence the recommendation was made that the number of OP applications to carrot crops should be limited to a maximum of three per year and alternatives to OP use sought.

Consumers meanwhile were reassured that peeling and 'topping' (removing the top 2–3 mm) carrots during preparation removed about 80% of residues. Relatively recent evidence suggests that there has been a rise in the percentage of samples exceeding MRLs in the UK, although the issue is complicated by the sampling and analysis of imported food products in combination with UK-produced food (CPA, 2000a; MAFF, 2000b).

Whilst residues in food are an important issue, it is the direct effect of pesticide use ins agroecosystems which causes environmental disruption. There are many important groups of beneficial and non-target organisms at risk from pesticide use and a single insecticide application can reduce total numbers in a cereal crop by up to 80%. Beneficial organisms affected include:

Pollinators Honey (*Apis mellifera*) and bumble bees are particularly important for the pollination of many fruits and vegetables, but are at risk when spraying occurs on or near to flowers where they are feeding (Williams, 1982; 1986). All of the reported and investigated incidents involving the poisoning of 'beneficial insects' by agricultural chemicals in 1999 in the UK involved honeybees. The majority of colonies were adversely affected by the organophosphate compound dimethoate (78%) whilst carbamate accounted for a further 12% (MAFF, 2000c).

Soil organisms May be adversely affected by fungicides, molluscicides and nematicides (RCEP, 1996; Indeherberg *et al.,* 1998). Applications of the fungicides metalaxyl and mancozeb can reduce populations of the beneficial mycoparasite *Pythium oligandrum* which acts as a natural control agent of many soil-borne pathogens. The carbendazim fungicides, including benomyl, are also particularly toxic to earthworm numbers. Herbicides are not believed to have a significant impact upon soil flora and fauna unless applied at very high rates (Conway and Pretty, 1991).

Beneficials The grain aphid (*Sitobion avenae*), a serious cereal crop pest in Europe has a huge variety of natural enemies that help control its populations. These natural enemies include linyphiid (money) spiders, carabid and staphylinid beetles, parasitoids, syrphids (hoverflies) and coccinellids (ladybirds) all of which are highly susceptible to the action of broad-spectrum insecticides. The loss of these natural predators can lead to the resurgence of pests which in turn incurs extra costs for the farmer through increased pesticide use and reduced yields.

Additionally, pesticides may also affect other non-target organisms. The UK Wildlife Incident Investigation Scheme (WIIS) investigates incidents of poisoning of domestic or wild animals (Prosser, 2001) and although the number of cases that occur are very difficult to estimate (Hart, 1990; Kjolholt, 1990) it is generally accepted that little direct lethal effect upon birds and small mammals occurs. Mortality due to the effects of slug pellets (upon small mammals) and seeds dressed with insecticide (on birds) has been recorded (Conway and Pretty, 1991). About 90% of all arable seed crops are treated with pesticide before planting; grainivorous birds such as the woodpigeon and house sparrow as well as soft-billed, normally insectivorous species, such as robins and dunnocks are at risk (MAFF, 2000b; MAFF, 2001; Prosser, 2001). Poisoning of birds may also occur by secondary poisoning. Raptors such as buzzards and red kites as well as owls are vulnerable to the effects of anticoagulant rodenticides (MAFF, 2000c).

Low level or 'sub-lethal' exposure are also apparent in non-farmed species via contaminated food sources and inhalatory or dermal exposure. Kjolholt (1990) reported patterns of altered behaviour amongst hares that would normally be associated with pesticide poisoning (although no detectable residues could be found in the animals) whilst Cilgi and Jepson (1995) investigated the susceptibility of two butterflies, *Pieris rapae* and *P. brassicae* to deltamethrin. When deltamethrin was applied topically to these insects, feeding and growth inhibition was demonstrated at sub-lethal doses. Residual exposure bioassays detected toxic effects at 0.002% of field application rates; a comparison of these results to rates of drift likely to occur suggested high levels of risk to larvae.

A further effect that pesticides may have upon wildlife is through the reduction in quality of their food or shelter, particularly in relation to herbicide use. For example, the local destruction of birds-foot trefoil from arable land in the UK has resulted in the elimination of associated populations of the common blue butterfly (*Polyommatus icarus*) because its larvae depend on this plant as a food source (Rands and Sotherton, 1986). Herbicide spray drift, as well as direct application, can also adversely affect the ecology of semi-natural habitats adjacent to farmland (MAFF, 2000c). One such habitat characteristic of European farmland is the hedgerow, traditionally associated with high floral and faunal diversity. It is well established that hedgerows not only raise insect diversity at their base, but also in the nearby parts of adjacent fields (Lewis, 1969). They are particularly important for harbouring over-wintering beneficial species such as beetles (Pollard, 1986) as well as providing nest sites and refuges for birds and food sources for other animals such as butterflies and bats. Herbicide drift or the application of herbicides directly to the hedge base reduces floral diversity and leads to species-poor plant communities dominated by pernicious weeds which offer little food or shelter to wildlife (Boatman *et al.*, 1999).

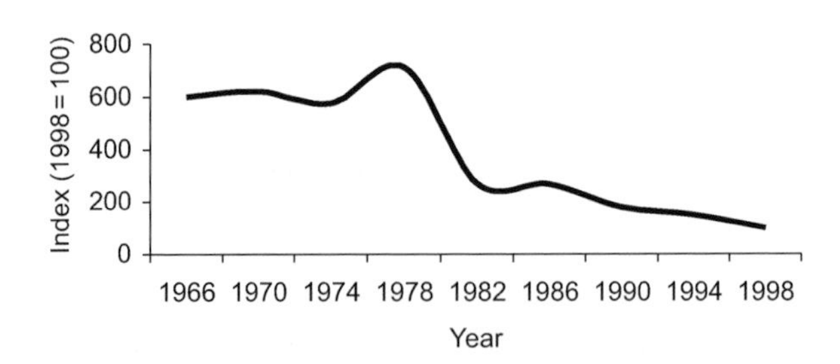

Figure 7.3 Numbers of grey partridges recorded by the British Trust for Ornithology from the Common Bird Census, 1966 to 1998 (Baillie *et al.*, 2001).

An example of the adverse indirect effect of pesticides is the serious decline in farmland bird numbers observed in the UK over the last 25 years or more (MAFF, 2000a). Insecticides account for only 3% by volume of all crop protection products, but they have a considerable impact upon bird populations (CPA, 2000a). In an examination of the trends in farmland bird numbers and their associated diet and ecology, it was concluded that the indirect effects of pesticide use contributed the decline of skylark, lapwing, swallow, blackbird, song thrush, starling and linnet (MAFF, 2001). Enough is known about the grey partridge (Figure 7.3) to conclude that pesticides have played a key role in its decline.

Numbers of grey partridge began to decline in 1952 and eventually fell to less than 20% of pre-1952 population densities. This decline has been attributed convincingly to a reduction in the number of insects in arable crops such as aphids and sawfly larvae which are an essential component of partridge chicks diet (Boatman *et al.,* 1999; Potts, 1986). The decline in insect food sources was due to the increased use of insecticides and the indirect effects of fungicides and herbicides. Since many insects depend upon weeds and fungi for food, insect numbers also decline as weeds and fungi are destroyed in crops.

7.10 PRACTICAL SOLUTIONS

The minimisation of environmental effects is being achieved by the adoption of a range of approaches that are being used as practical solutions to solve the problem of pesticide pollution.

Good practice

The encouragement of good agricultural practice (GAP) is a well-established approach to reducing environmental pollution from modern farming systems (Jordan, 1993) and is enshrined, to varying degrees, within national and international legislation. The 1991 EC Pesticide Registration Directive requires that pesticides are to be used 'properly' in compliance with label conditions and good agricultural practice, including '...the principles of integrated control whenever possible' (Beaumont, 1992).

The UK government has published guidelines on pesticide usage which are intended both to minimise their environmental impact and allow safe usage. The Code of Practice for the Safe Use of Pesticides on Farms and Holdings was prepared jointly by MAFF (now DEFRA), the Health and Safety Commission and the Department of the Environment, Transport and the Regions (MAFF, 1998e). Consequently, it is a 'combined' code of practice with legislative status under both the Food and Environment Protection Act 1985 (with its Control of Pesticides Regulations 1986) and the Health and Safety at Work etc. Act 1974 (with its Control of Hazardous Substances to Health Regulations 1988).

The Code of Practice gives practical guidance to farmers, commercial growers and to those who give advice or practical assistance, such as agricultural contractors. Different aspects of using of pesticides are dealt with according to the particular perspective of each user group in order to enable them to fulfil their legal obligations fully. Some of the main aspects of pesticide use relating to farmers are outlined below.

User training

It is a legal requirement in the UK that anyone who uses pesticides on a farm must have 'adequate instruction, training and guidance in the safe, efficient and humane use of pesticides and be competent for the duties, which they are called upon to perform'. Training is required for the proper use of equipment and safe working practices and this is widely seen as a key to limiting the potentially adverse effects of pesticide residues on non-target organisms and environments (CPA, 2000b). The training is specific to the type of equipment being used and also includes other aspects including the preparation of the pesticide prior to application, the hazards and risks involved. Employers have responsibility for ensuring that operators are given 'information, instruction, training and guidance... and know the risks to health created by exposure to pesticides, and the precautions which should be taken'.

Planning and preparation

The Code of Practice emphasises that farmers should firstly be certain that a pesticide application is required. More than 90% of all crop protection treatments applied in the UK, such as pesticides, are now applied following advice from qualified agronomists (CPA, 2000b). Prior to application, other control measures should be considered such as biological control or integrated pest management. Informed decision-making must be based on correctly identifying the pest, previous experience of the problem, selection of an appropriate pesticide and ensuring effective application.

Table 7.8 Examples of special precautions needed by pesticide applications methods (MAFF, 1998e).

Application method	Special precautions
Hand-held sprayer/ reduced-volume spraying	Prevent unwanted drift and contamination of user. Use appropriate nozzle, correct boom height and avoid drift.
Fog, mist and smoke and space treatments	Ensure areas are not occupied and exclusion notices displayed. Assess environmental risks by escape.
Granule and bait application	Special attention needed to dusts and vapours associated with some granules.
Seed treatments	Care needed as some products leave dusty residue in containers. Care needed when cleaning seed drill equipment or clearing spilt dressed seed.
Dipping and drenching plant treatments	Ensure treated material is labelled properly if unattended between treatment, planting or storage.

The product chosen must be appropriate, and the code states that:

- it should be approved for the intended use and situation;
- it can be safely prepared and applied by the operators;
- it can be applied in time to leave any required interval before harvest;
- it presents the least risks to health while still achieving effective pest control;
- it does not present undue risk to livestock and the environment, for example to bees and fish, creatures that are susceptible to certain pesticides.

Pesticides may be applied in a number of ways and each method requires special precautions (Table 7.8). Prior to application, a series of checks must be conducted to ensure that: the application equipment is functioning correctly, the pesticide being used is correct; the environment is protected, emergency provision has been planned and that disposal of containers, etc. has been considered.

Working with pesticides

Problems with pesticides often occur during application (CPA, 2000b). The Code of Practice includes advice on the use of control measures, identification of dangerous practices, pesticide container handling and precautions needed when filling equipment. In addition, special consideration should be given to the protection of particularly sensitive animals such as honeybees (Box 7.4), the prevention of pesticide drift (particularly a problem in moderate or gentle winds (Table 7.9), and the protection of neighbours and walkers. Boundary areas, water courses and surface drainage channels, such as ditches, should not be sprayed and must be protected from drift, possibly by leaving an unsprayed strip of crop at the field margin to prevent drift out of the area. Exposure monitoring should be done after use and in addition, maintenance requirements of equipment and health surveillance are identified.

Box 7.4 The effect of pesticides on honeybees

The code of practice for the safe use of pesticides gives special advice for the protection of these economically important non-target insects and also identifies the need to liase with local beekeepers. The Code states that the following precautions should be taken:

DO NOT
Use pesticides harmful, dangerous or high risk to bees if crops or weeds are in open flower or part bloom unless this is allowed by the product label.
Let pesticide drift into beehive entrances (where it can kill brood and adult bees) or into hedgerows or fields where bees may be foraging.
Spray unless you have to.

DO
Follow closely the environmental protection instructions on the product label, and the guidance in the Code.
Use a product that presents the least hazards to bees.
Check for bees foraging or visiting plants.
Spray in the evening when bees have stopped flying, as this allows several hours for the pesticide to dry before bees become active.
If you must spray during the day, choose a cool cloudy one, or the early morning.

Table 7.9 Wind speed conditions and the suitability of pesticide spraying (MAFF, 1998e). High temperatures combined with low humidity reduce the size of spray droplets by evaporation, so increasing the risk of spray drift.

Air speed at boom height (km h^{-1})	Beaufort scale at 10 m	Description	Spraying conditions
<2.0	Force 0	Calm	Use a medium or coarse spray
2.0–3.2	Force 1	Light air	Acceptable conditions
3.2–6.5	Force 2	Light breeze	Ideal conditions
6.5–9.6	Force 3	Gentle breeze	Increased risk of spray drift
9.6–14.5	Force 4	Moderate breeze	Spraying not recommended

Disposal of waste

The Code provides guidance on handling and disposal of waste pesticides and gives advice on waste minimisation, tank washing, disposal of dilute and concentrate wastes, disposal of containers, waste packaging and contaminated material. Waste minimisation is identified as a key to reducing disposal problems.

Keeping records

The Code states that good records should be kept involving the storage, application and disposal of pesticides. The code identifies the importance of recording 'accidental contamination of people, honey bees, other creatures, land, water or non-target crops'.

Other considerations

In addition to the Pesticide Code of Practice, the Codes of Good Agricultural Practice for the Protection of Soil, Water and Air (MAFF, 1998a,b,c,d) make the following recommendations:

Water This code gives advice on pesticide storage and specifies that stores should not be built where there is a risk of polluting watercourses or groundwater. The guide refers the reader to Health and Safety Executive guidance note CS19 'Storage of Approved Pesticides: Guidance for Farmers and other Professional Users' (British Standard BS5502: Part 81) for advice on the construction of stores. If spillages occur, the code states that quick action is required, including notification of the appropriate statutory agency. Pesticides should never be applied when there could be drift onto waters unless specific approval is given. The code also gives guidance on disposal of waste pesticides (concentrated and dilute), disposal of containers and contaminated material.

Soil This code states that most pesticides are organic in nature and are broken down so they have little long-term effect on soils. By law, these compounds must only be applied at a rate and in the way specified in their authorisation. Some pesticides based on inorganic materials, such as copper, may also cause contamination of land and special care is needed.

Air This code states that although burning of containers contaminated by pesticides may be permitted under certain circumstances, that 'there are few circumstances where other practicable methods of disposal cannot be found' and that these disposal methods are recommended.

Other *ad hoc* initiatives have also been developed in the UK to encourage good practice to minimise the risk of pesticide pollution. For example, the Crop Protection Association runs training programs on environmental/product information for those advising on pesticide usage (CPA, 2000b).

7.11 NEW TECHNOLOGY

Reductions in pesticide usage and increases in cost effectiveness can be achieved through relatively simple changes in application technology. The basic components of a pesticide sprayer are a tank, pump, boom and nozzles. The pesticide in solution is pumped under pressure from the tank to the nozzles where it is commonly discharged as a fine atomised spray. The main problem with the use of conventional 'hydraulic' nozzles is that they produce a wide range of droplet sizes (10–700 μm); those droplets that are very small are particularly susceptible to drift, whereas those that are very large are prone to bounce, or run off the crop surface.

Alternative designs of nozzle can overcome this problem by producing a narrower, more optimal, range of droplet sizes that increase the efficiency of application (Culpin, 1992; CPA, 2000b). This reduces both the volume of pesticide required and the risk of pollution. Alternative types of nozzle which have been developed include:

Vibrated Jet Distribution Liquid is delivered at low pressure through holes in an oscillating cylinder operated by a tiny electric motor. This produces a band of spray that is free of fine droplets and therefore not susceptible to drift;

Controlled Droplet Application (CDA) Liquid drops onto a spinning disc and is forced to the edge of the disc where it is atomised by a serrated surface. This produces a very consistent droplet size that can be precisely controlled to reduce drift by varying flow rate and/or the speed of the spinning disc;

Electrostatic Nozzles Very low volumes of liquid are atomised to form tiny droplets that are electrostatically-charged and therefore attracted directly to the crop surface without risk of drift.

Inefficiency of nozzles caused through blockages or the usage of sub-standard equipment can also cause environmental problems. This is being tackled by a range of agricultural contractor and machinery associations promoting testing, validation and certification of equipment in order that at least 30% of all eligible users have been checked by 2003 (CPA, 2000b).

The majority of modern tractor-mounted, trailed or self-propelled sprayers are also fitted with automatic application rate control systems. These are based upon a computer that uses forward speed and fluid flow information from sensors to correct the rate of spray application and eliminate errors as the sprayer varies in forward speed. For example, over-application of pesticide due to the sprayer losing forward speed on a slope can be avoided. Stafford and Miller (1993) describe attempts to use similar technology to ensure that herbicide is only applied to detected weed patches in arable crops – a 'precision agriculture' approach. Precision farming aims to accurately record and 'map' the variability of individual fields (Terry, 1995; Glendinning, 1999) thus facilitating the controlled application of agrochemicals.

Adjuvants

Another approach to improving the efficiency of application, and therefore environmental safety, of pesticide use is to add adjuvants, inert chemical substances that affect the physical properties of a pesticide spray (Makepeace, 1996). There are many different types of adjuvant; some are synthetic chemicals (e.g. non-ionic wetters), others are naturally-derived (e.g. emulsified rape oil). They may improve the environmental and agronomic behaviour of pesticides in a number of different ways:

- protecting against poor weather, especially cold damp conditions;
- reducing losses due to rain washing the pesticide off its target;
- reducing drift by increasing spray droplet size;
- increasing retention of the pesticide on the plant;
- assisting plant uptake of pesticides;
- increasing chemical activity of the pesticide.

Codacide oil, for example, is an emulsified vegetable oil product that improves spray droplet size and therefore greatly reduces wastage due to drift. Consequently, significant financial savings (particularly with some herbicides) can be made whilst reducing environmental impact and without compromising effective weed control. Since codacide oil is vegetable-based, it is also biodegradable.

New formulations

One way of reducing the environmental impact of compounds used as pesticides is to make them 'safer' by using those that are specific to their target pest and less persistent. Examples of specific pesticides include the systemic carbamate insecticide pirimicarb, which has relatively low mammalian toxicity and high biochemical selectivity for aphids.

The approval process for the use of new pesticides is gained in the UK through the Control of Pesticides Regulations 1986 or the Plant Protection Products Regulations 1995 (which implements Directive 91/414/EEC). The full evaluation of a new product an extensive and costly process which all new products must be subjected to (CPA, 2000a).

7.12 ALTERNATIVES TO SYNTHETIC PESTICIDES

Naturally derived pesticides

The use of natural plant-based pesticides is an alternative to those that are synthetically derived. Although compounds such as derris and pyrethrum display relatively low environmental persistence, they are broad-spectrum in action and may pose a risk to a number of organisms. Derris and pyrethrum are, for example, highly toxic to fish (Worthing, 1991). Conway and Pretty (1991) identify antifeedants that render plants unattractive and unpalatable to pests as promising plant-based compounds. For example, chemical derivatives of neem (*Azadirachta indica*), fenugreek (*Trigonella*) and turmeric (*Curcuma longa*) are all strong repellents to herbivorous insects.

Microbial pesticides

Equally promising in terms of selectivity and minimal environmental pollution are the microbial pesticides. Micro-organisms that might be used include bacteria, fungi, viruses and protozoa (Lisansky *et al.*, 1991). With the development of new processing and formulation methods, plus the use of genetic engineering techniques to enhance the pesticidal potency of individual micro-organisms, it seems likely that the availability and use of microbial pesticides will become increasingly widespread. One of the most successful microbial pesticides to-date is *Bacillus thuringiensis* (Bt). This was discovered around 1901 and was first available commercially in the 1930s for use against caterpillars (Lisansky *et al.*, 1991). Other strains of Bt are also now available for use against flies and beetles. Bt is commonly formulated as a wettable powder for mixing with water and application as a spray. To work effectively, bacteria from the formulated product must be ingested by the pest. Once inside the gut the bacteria produce a crystalline compound containing a variety of toxic proteins called 'δ-endotoxins' which cause paralysis and disruption of feeding.

Certain fungi are adapted to live on insects and hence show potential as microbial pesticides. A good example is *Verticillium lecanii*, a pathogenic fungus that naturally infects aphids and scale insects in the tropics and sub-tropics (Lisansky *et al.*, 1991). *V. lecanii* has been commercially-available in the UK since the early 1980s for the control of the glasshouse whitefly (*Trialeurodes vaporariorum*), a common and troublesome pest in protected crops.

V. lecanii spores are normally applied by the grower as a conventional high volume spray. The spores germinate on the whitefly pests and infection occurs by the penetration of fungal growth through the cuticle of the insect. Successful parasitism is observed as white 'fluffy' insect bodies some 6–12 days after application of the spores. Under favourable conditions, more spores can be produced and spread from these infected insects leading to the suppression of the pest population for several weeks (Helyer and Richardson, 1991). The main disadvantage of *V. lecanii* is the high degree of skill and environmental control required ensuring its use is fully effective. In particular, the fungus requires a sustained period of high relative humidity after spore application to ensure successful germination, infection and further spore spread. Until fungi have been developed which have greater tolerance to adverse environmental conditions and display more rapid infection cycles, it is likely that their full potential as microbial pesticides, at least in temperate agricultural systems, will remain limited (Lisansky *et al.*, 1991).

Non-chemical weed control

Prior to the widespread availability and use of herbicides, an important strategy for controlling weed populations was the use of crop rotations to produce a diverse agronomic environment. The adoption of different sowing dates, crop growth periods and cultivation operations reduced the likelihood of weeds such as wild oats and blackgrass from establishing and becoming an agronomic problem (Lampkin, 1990). Furthermore, timing of cultivation can have a considerable bearing on diseases such as barley yellow dwarf virus, which can be eliminated if drilling is delayed until early October (MAFF, 1999). Whilst the reintroduction of a traditional rotational cropping system may not be viable on all farms, other non-chemical weed control techniques may be adopted. Alternative husbandry practices include cultivation, increased sowing rates and direct thermal and mechanical intervention.

Appropriate cultivation Prior to the establishment of a crop can play a useful preventative role in weed control (MAFF, 1999). Crop sequences that alternate cereals with broad-leaved crops allow grass weeds to be targeted in the broad-leaved crops and broad-leaved weeds in the cereals (MAFF, 1999). This can be undertaken in combination with the preparation of a series of 'stale' or 'false' seedbeds before crop establishment is due. Weed seeds brought to the surface during cultivation are encouraged to germinate and grow in the seedbed for about 10 days before being destroyed by further cultivation. Sufficient time between cultivations to allow weed germination is required for this approach to be effective.

Night-time cultivation May be adopted to reduce weed germination. The technique exploits the fact that many weed seeds are photoblastic, requiring exposure to daylight to induce germination. It is possible to minimise the daylight exposure of those dormant seeds buried deeply below the soil surface by performing cultivation operations in darkness or at sufficiently low light levels. On-farm trials in Germany showed a marked reduction in weed coverage from 80% to 2% when crops were sown at night rather than at mid-day (Hartmann and Nezadal, 1990). Those weeds showing dramatic reduction in population when switching from daytime to night-time cultivation included wild chamomile (*Matricaria chamomilla*), small toadflax (*Chaenarrhinum minus*), cleavers (*Galium aparine*), chickweed (*Stellaria media*) and fat hen (*Chenopodium album*). Despite attracting considerable interest, the use of night-time cultivations for weed control have not been successfully applied in the UK (Samuel, 1992).

Thermal intervention There have been many advances in non-chemical weed control over the last 15–20 years, both Parish (1990) and Lampkin (1990) provide useful reviews of the range of equipment now available for controlling weeds by direct thermal (and mechanical) intervention in cereals and row crops. Most thermal weeders are tractor-mounted and use liquefied petroleum gas (LPG) as fuel. The weeders available are becoming increasingly sophisticated and can be used for pre- or post-emergence weed control. The principle of thermal weed control is not to burn off the weeds, but to apply sufficient heat to severely damage the plant cells so that the weed will wither and die. This generally involves raising the plant tissue to a temperature of 100 °C for 0.1 seconds. Two basic designs of flame weeder are available: the 'flame contact' type and the 'infra-red type' involving a radiant surface. The 'flame contact' type is cheaper to buy and can be more effective than the 'infra-red' type, but is less economical on fuel. Crops such as carrots, which have a long germination period during which weed seedlings emerge before the crop can be very effectively treated by pre-emergence thermal weeding although the timing of the thermal treatment is crucial for maximum weed control and minimum crop damage. Other crops will, at specific growth stages, tolerate post-emergence thermal treatment although crop protectors and shields may be needed.

Mechanical intervention Inter-row cultivation is a common form of mechanical weed control in row crops and there are a number of designs of tractor-mounted cultivators available. These vary from traditional spring-tine cultivators to novel PTO-powered brush-type machines. Mechanical weed control using a comb-type harrow can also be very effective in (non-row) cereal crops provided that the crop is sufficiently well-established to avoid damage whilst only harbouring immature weeds which can be easily disturbed.

Non-chemical pest control

Crop management practices, such as the use of resistant varieties or varietal mixes are well-established for the cultural control of insects and diseases. Many of these techniques have remained integrated within modern pest control strategies, although their significance declined as the cost-effectiveness of pesticides encouraged farmers to rely upon chemical control. Interest in these strategies is now increasing as balanced approaches to pest and disease control encompassing a broad range of control measures are sought.

An example of a move to integrate these methods is in carrot crops in the UK. The production of high quality, marketable carrots in the UK is dependent upon the effective control of the carrot fly (*Psila rosae*), the larvae of which frequently cause serious damage to carrot crops by burrowing into the developing roots. The use of OP insecticides to control carrot fly has increased significantly (Box 7.3). A number of cultural control methods to facilitate the production of high quality carrots with minimal use of insecticides are therefore under development. These include (PSD, 1995):

Resistant varieties Varieties with some resistance to carrot fly have been developed and can reduce the level of damage by up to 50%, but will not prevent the occurrence of significant and commercially unacceptable damage.

Crop covers Floating film crop covers can be used to exclude a number of aerial pests, including carrot fly. However, using such covers is costly, their durability can be questionable, they may reduce yield and are difficult to integrate with other chemical controls.

Spatial separation The carrot fly is not a strong flier and so the risk and intensity of damage can be reduced by minimising their 'carry-over' from one field to another. Main crop carrots especially (because they remain in the field for longer) should be sited at least 2 km from both previously infested fields and early carrot crops grown in the same season, although this may not always be practicable.

Harvesting strategy With the demand for very high quality carrots in the UK, current harvesting practice is to leave main crop carrots in the ground during winter in order to avoid the risk of blemishes and skin 'silvering' that occurs when lifted early and put into cold store. However, since the majority of late generation carrot fly damage occurs after October, this practice of 'field storage' contributes to the need for high levels of chemical control to reduce damage to commercially acceptable levels.

Lifting carrots in mid to late October can significantly reduce the risk of carrot fly damage and therefore enable a reduction in pesticide use. However, storage costs are greatly increased and there remains opposition to the practice from the supermarkets. A further option being explored is the development of an integrated crop management strategy for carrots which aims to balance modifications in harvesting practice with continued pesticide application. A leading example of this is the NFU-Retailer Integrated Crop Management Protocol for Fresh Carrots developed jointly in the UK by the National Farmers Union, representative supermarkets and members of the Carrot Marketing Board.

Agricultural crop pests are attacked by a wide range of naturally-occurring enemies, including parasitic wasps and flies, ladybirds, spiders and hoverflies, all of which have potential for exploitation as alternatives to pesticide use (Box 7.5). The aim of pest control using natural enemies is not to eradicate the pest population, but to reduce it to an acceptable and economically viable level and maintain it there. By comparison with pesticide use, the pest population may be larger and it may fluctuate more, but control tends to be cheaper and more durable (Conway and Pretty, 1991).

Box 7.5 Common pests in horticulture and their biological control agents

Whitefly

Whitefly (*Trialeurodes vaporariorum*) is a familiar pest attacking a wide range of protected crops. The adult is 1–2 mm long, winged and white/creamy yellow in colour (resembling a small white moth). The adult whitefly are usually found at rest under leaves, where the females lay their eggs. The eggs hatch to produce mobile 'crawlers', which attach themselves to the leaf for feeding before developing into static, white, waxy 'scales' (the main feeding and growing stage). The 'scales' pupate and the emerging adults move towards to the top of the plant to feed on the younger, soft foliage and to lay eggs. Low levels of infestation may not produce any visible signs of damage. At high levels leaves and stems will be covered with 'scales'. These draw sap from the plant, reducing crop vigour and distorting young growing tips. They also secrete minute droplets of sugary 'honeydew', which encourages the growth of sooty moulds leading to reduced photosynthetic ability and disfigured produce. Additionally, both the adults and the 'scales' are potential virus carriers. Whitefly can be effectively controlled by *Encarsia formosa*, the most widely used biological control agent in horticulture. *Encarsia* is a parasitic wasp introduced to the crop in the form of parasitised whitefly 'scales' which are commonly supplied loose with a dispenser, on fresh leaf material or on cards.

Red spider mite

The red spider mite (*Tetranychus urticae*) is one of the most damaging pests that attack protected crops. Only the over-wintering adult female is actually red, the summer form is green or yellow-brown. The mites hibernate in cracks and crevices of the greenhouse or crop debris, and usually emerge in spring as the temperature and day length increase. Eggs are laid under the leaf and hatch to reveal white, 6-legged larvae. Feeding spider mites cause a fine, yellowish speckling on upper leaf surfaces (initially at the base of the plant) and, if left unchecked, will cause leaves to shrivel and may eventually kill the plant. The mites also spin silken webs and at high levels of mite infestation these may cover much of the plant, stunting new growth. Red spider mite can be controlled with a predatory mite, *Phytoseiulus persimilis*, which has been successfully used in the greenhouse industry since the early 1960s.

The use of natural enemies for pest control is usually referred to as biological control. There is an important distinction between 'classical' biological control involving the direct release of new or exotic natural enemies into the crop, and the encouragement of existing natural enemy populations by enhancing the environment in which they live.

Stimulated principally by problems of pesticide resistance among a number of major pests, classical biological control methods have been used very successfully in UK glasshouse crops, such as tomatoes and cucumbers, for some 70 years. Consequently, there are now a large number of natural pest predators commercially available for use in protected cropping. The best known examples (Box 7.5) are the whitefly parasite (*Encarsia formosa*) and the spider mite predator (*Phytoseiulus persimilis*) which are used on up to 50% of protected tomato crops and up to 80% of protected cucumber crops in the UK (Helyer and Richardson, 1991).

Many other natural enemies for controlling pests of protected crops have now been evaluated and are in commercial mass production, including an aphid predator (*Aphidoletes aphidimyza*), mealybug predator (*Cryptolaemus montrouzieri*) and a voracious polyphagous predator, the green lacewing (*Chrysoperla carnea*). Helyer and Richardson (1991) summarised the advantages of biological control in the glasshouse, compared to purely chemical control, as environmental, increased yield and quality and minimisation of resistance. The advantage of releasing natural enemies into a glasshouse is that they are contained in a favourable environment which can easily be controlled to ensure that the introduced organisms survive and thrive to have the desired effect upon the pest population.

Applying classical biological control methods to outdoor field-scale crops is not so easy, although there have been some notable successes. *Trichogramma*, for example, has been used to control outbreaks of the European corn borer, *Ostrinia nubilalis* on maize and has been used successfully in France, Switzerland and Germany. The parasitoids are usually reared on the eggs of the flour moth (*Ephestia* spp.) and released in high numbers; up to 600,000 parasites ha^{-1} three to four times during the egg-laying period of the corn borer. Release is determined from pheromone catches of adult moths and from surveys to monitor the appearance of corn borer eggs in the field (Burn, 1987).

An alternative approach to exploiting natural enemies for pest control in field-scale agricultural crops is to enhance the environment in which they live. In particular, populations of natural enemies of a given agroecosystem can be encouraged by removing adverse factors (e.g. broad-spectrum insecticides) as much as possible and increasing the diversity of the agroecosystem and its neighbouring semi-natural environment by fostering crops and wild plants that favour natural enemies.

There are many examples of the practical application of these principles to field crops and two are given below which relate specifically to the control of aphids. Aphids are important pests on a wide variety of agricultural and horticultural crops, causing both direct damage through their feeding and indirect damage by spreading virus diseases. However, although aphids reproduce at a very high rate their numbers are often checked by an array of natural enemies. These fall into two main groups, general (polyphagous) predators and aphid specific predators, both of which have an important role to play in aphid control. Each group also has its own requirements in terms of a suitable environment and an understanding of these requirements can be used to enhance predator numbers via the use of specific management practices and strategies.

Polyphagous predators, including ground and rove beetles (e.g. *Agonum dorsale*, *Tachyporus hypnorum* and *Demetrias atricapillus*) and linyphiid spiders, are of particular importance for aphid control early in the season as they are already in the crop living on other sources of food when the aphids arrive. They therefore have the potential to check an invading aphid population before it becomes excessive. A number of the polyphagous predators overwinter in field boundaries and in spring migrate from the hedges and field margins into the crop to begin searching for food. Although some have wings, most polyphagous predators do not and may take several weeks to penetrate into the centre of large fields. During this time, aphid levels in the crop may be increasing to the level of an outbreak. Researchers in the UK have investigated providing suitable overwintering sites in the centre of fields through the use of mid-field ridges or 'beetle banks' (Box 7.6), in an attempt to encourage beetles and spiders to remain and be ready for rapid dispersal in the spring (Harwood *et al.*, 1992).

Box 7.6 Establishing and looking after mid-field ridges

Mid-field ridges, or 'beetle banks' are a simple way of encouraging predatory insects to assist aphid control and the rationalisation of insecticide use. Advice on their creation and maintenance is provided by The Game Conservancy Trust (1992). Each ridge is a bank of earth approximately 0.4m high and 1.5–2.0 m wide. It is created during normal cultivations by the careful two-directional ploughing of opposing furrows. The number of ridges formed depends upon the size of the field and the quality of its existing boundaries. A 20-hectare field which has established boundaries with raised underbanks and an abundance of dense tussocky grasses, such as cocksfoot (*Dactylis glomerata*) and Yorkshire fog (*Holcus lanatus*), will probably only need one ridge across its centre. A similar sized field with poor boundaries, no underbank and few tussocky grasses may require two ridges dividing the field into three equal parts. Once established, the ridge should be drilled in the spring with a 50:50 mix of Yorkshire fog and cocksfoot at a rate of 3–4 $g\,m^{-2}$. As the grasses grow they should exclude most weeds, but may themselves be susceptible to herbicides sprayed on the adjacent crop. A 1-m wide strip between the ridge and the crop should minimise this problem. During the summer the ridges can be mown or 'topped' to prevent excessive growth and encourage the formation of tussocks. After 2 or 3 years, the ridges should harbour sufficient spiders and beetles to have significant effect upon crop pests. If farming practices change, the ridges can be removed and re-created elsewhere.

Aphid specific predators, such as hoverflies, ladybirds and lacewings, will not enter the crop until there is food for them. However, once in the crop they are more efficient than polyphagous predators at controlling aphid numbers. Some of the most important aphid specific predators are small parasitic wasps (Powell, 1996). There are around 5500 species of parasitic wasp in the UK, of which approximately 75 are aphid specific. These can have a big impact upon aphid populations, but their activity must again be synchronised with the increase in aphid numbes early in the growing season. Unfortunately, this early season synchronisation does not always happen as the wasps are dispersed to a range of semi-natural habitats for over-wintering and may not return to the crop in sufficient time to curb an aphid attack. This delay between aphid infestation and wasp arrival can be reduced by encouraging the wasps to overwinter adjacent to the crop and by ensuring conditions are favourable for their activity in the spring (Box 7.7).

Box 7.7 Encouraging parasitic wasps for aphid control

Powell (1996) suggests the following guidelines for encouraging parasitic wasps:

Leave some areas of natural vegetation uncut
Parasitic wasps rely upon areas of natural and semi-natural as over-wintering sites and these should be maintained on the farm as much as possible (e.g. areas of permanent grassland, either in pastures or field margins).

Maintain a diversity of habitats
When the aphids disappear from crop fields in late summer, parasitic wasps need to find alternative aphid hosts. A range of aphids can be maintained on the farm in order to support the wasps 'out-of-season' (especially early spring and autumn) by creating and/or encouraging a variety of crop and semi-natural habitats.

Avoid insecticide sprays whenever possible, especially early in the spring
Parasitic wasps are both killed and repelled by insecticides. Even the use of a selective aphicide will also have a detrimental effect upon wasp populations because of: (a) the loss of potential hosts; and (b) the destruction of parasite larvae already developing inside aphids.

REFERENCES

Aislabie, J.M., Richards, N.K. and Boul, H. (1997) Microbial degradation of DDT and its residues – a review. *New Zealand Journal of Agricultural Research* **40**, 269–282.

Alloway, B.J. (1995) *Heavy Metals in Soils.* 2nd Edition. Blackie Academic and Professional, London.

Alloway, B.J. and Ayres, D. (1993) *Chemical Principles of Environmental Pollution.* Blackie Academic and Professional, London.

Baillie, S.R., Crick, H.Q.P., Balmer, D.E., Bashford, R.I., Beaven, L.P., Freeman, S.N., Marchant, J.H., Noble, D.G., Raven, M.J., Siriwardena, G.M., Thewlis, R. and Wernham, C.V. (2001) Breeding Birds in the Wider Countryside: their conservation status 2000. BTO Research Report No. 252. BTO, Thetford. URL: http://www.bto.org/birdtrends

Baird, C. (1995) *Environmental Chemistry.* W.H. Freeman, New York.

Barber, S.A. (1995) *Soil Nutrient Bioavailability: a Mechanistic Approach.* 2nd Edition. Wiley, New York.

Bentabol, A. and Jodral, M. (1995) Occurrence of organochlorine agrochemical residues in Spanish cheeses. *Pesticide Science* **44**, 177–182.

Beaumont, P. (1992) Pesticides, Europe and the environment: a review. *Pesticide News* **16**, 12–15.

Bigler, F., Forrer, H.R. and Fried, P.M. (1992) Integrated crop protection and biological controls in cereals in Western Europe. In: *Biological Control and Integrated Crop Protection: Towards Environmentally-Safer Agriculture* (Eds J.C. van Lanteren, A.K. Minks and O.M.B. de Ponti), Pudoc Scientific Publishers, Wageningen, pp. 95–116.

Boatman, N., Stoate, C., Gooch, R., Rio Carvalho, C., Borralha, R., de Snoo, G. and Eden, P. (1999) The environmental impact of arable crop production in the European Union: Practical options for improvement. Allerton Research and Education Trust, UK, Study Contract B4-3040/98/000703/MAR/D1.

Brown, C.D., Hodgkinson, R.A., Rose, D.A., Syers, J.K. and Wilcockson, S.J. (1995) Movement of pesticides to surface waters from a heavy clay soil. *Pesticide Science* **43**, 131–140.

Burn, A.J. (1987) Cereal Crops. In: *Integrated Pest Management* (Eds A.J. Burn, T.H. Coaker and P.J. Jepson). Academic Press, London, pp. 89–112.

Carson, R. (1985) *Silent Spring.* Penguin, Harmondsworth (first published in 1962 by Houghton Mifflin).

Cilgi, T. and Jepson, P.C. (1995) The risks posed by deltamethrin drift to hedgerow butterflies. *Environmental Pollution* **87**, 1–9.

Cleemann, M., Poulsen, M.E. and Hilbert, G. (1995) Deposition of lindane in Denmark. *Chemosphere* **30**, 2039–2049.

Conway, G.R. and Pretty, J. N. (1991) *Unwelcome Harvest: Agriculture and Pollution.* Earthscan, London.

CPA (2000a) Crop Protection Association Handbook 2000. Crop Protection Association, Peterborough.

CPA (2000b) Minimising the environmental impacts of crop protection chemicals. Revised proposals. Crop Protection Association, Peterborough.

Cremlyn, R. (1979) *Pesticides: Preparation and Mode of Action.* John Wiley and Sons, Chichester.

Culpin, C. (1992) *Farm Machinery.* Blackwell Scientific Publications, Oxford.

DoE/MAFF (1995) Rural England: A nation committed to a living countryside (Cm 3016). HMSO, London.

Environment Agency (2000) *Pesticides 1998. A summary of monitoring of the aquatic environment in England and Wales*. Environment Agency, Bristol.

Eurostat (1998) *Derived from data in crop protection: Half yearly statistics*. European Union, Brussels.

Filser, J., Fromm, H., Nagel, R.F. and Winter, K. (1995) Effects of previous intensive agricultural management on micro-organisms and the biodiversity of soil fauna. *Plant and Soil* **170**, 123–129.

Funari, E., Bottoni, P. and Giuliano, G. (1991) Groundwater contamination by herbicides – processes and evaluation criteria. In: *Chemistry, Agriculture and the Environment* (Ed. M.L. Richardson). Royal Society of Chemistry, Cambridge, pp. 235–254.

Game Conservancy (1992) Helping nature to control pests (information leaflet sponsored by Rhône-Poulenc). The Game Conservancy, Fordingbridge.

Gerstler, Z. (1991) Behaviour of organic agrochemicals in irrigated soils. In: Chemistry, *Agriculture and the Environment* (Ed. M.L. Richardson). Royal Society of Chemistry, Cambridge, pp. 332–369.

Glendinning, J.S. (1999) *Australian Soil Fertility Manual*. CSIRO Publishing, Collingwood, Australia.

Graham-Bryce, I.J. (1987) Chemical Methods. In: *Integrated Pest Management* (Eds A.J. Burn, T.H. Coaker, P.J. Jepson). Academic Press, London, pp. 113–160.

Harrad, S.J. (1996) The environmental behaviour of toxic organic chemicals. In: *Pollution, Causes, Effects and Control* (Ed. R.M. Harrison, 3rd Edition). The Royal Society of Chemistry, Cambridge, pp. 367–392.

Hart, A.D.M. (1990) The assessment of pesticide hazards to birds: the problem of variable effects. *Ibis* **132**, pp. 192–204.

Hartmann, K.M. and Nezadal, W. (1990) Photocontrol of weeds without herbicides. *Naturwissenschaften* **77**, 158–163.

Harwood, R., Hickman, J., MacLeod, A. and Rothery, F. (1992) Biological control in the field: a new approach. *New Farmer & Grower* **35**, 17–19.

Hassall, K.A. (1990) *The Biochemistry and Uses of Pesticides*. (2nd Edition). Macmillan Press, London.

Helling, B., Reinecke, S.A. and Reinecke, A.J. (2000) Effects of the fungicide copper oxychloride on the growth and reproduction of *Eisena fetida* (Oligochaeta). *Ecotoxocology and Environmental Safety* **46**, 108–116.

Helweg, A. (1994) Threats to water quality from pesticides – case histories from Denmark. *Pesticide Outlook* October, 12–18.

Helyer, N. and Richardson, P. (1991) *Biological control – protected crops*. Grower Digest 11, Grower Publications, London.

Herrling, B., Alesi, E.J., Bott-Breuning, G. and Diekmann (1993) In-situ aquifer remediation from volatile or biodegradable organic compounds, pesticides, and nitrate using UVB technique. In: *Contaminated Soil '93* (Eds F. Arendt, G.J. Annokkee, R. Bosman, R. and W.J. van den Brink). Kluwer Academic Publishers, Netherlands, pp. 1083–1092.

Hurst, P., Beaumont, P., Jorgensen, C.E. and Winther, S. (1992) *Pesticide Reduction Programmes in Denmark, the Netherlands and Sweden*. World Wide Fund for Nature International, Gland, Switzerland.

Indeherberg, M.B.M., De Vocht, A.J.P. and Van Gestel, C.A.M. (1998) Biological interactions: Effects on and the use of soil invertebrates in relation to soil contamination and in situ soil reclamation. In: *Metal-Contaminated Soils: In situ Inactivation and Phytorestoration* (Eds J. Vangronsveld and S.D. Cunningham). Springer-Verlag and R.G. Landes Company, Berlin, pp. 93–119.

Jordan, V.W.L. (1993) *Scientific basis for codes of good agricultural practice*. Report No. EUR 14957, Commission of the European Communities, Luxembourg.

Kjolholt, J. (1990) Distribution of pesticides and potential exposure of non-target organism following application. In: *Pesticide Effects on Terrestrial Wildlife* (Eds L. Somerville and C. Walker). Taylor & Francis, London, pp. 33–64.

Lampkin, N. (1990) *Organic Farming*. Farming Press, Ipswich.

Lewis, T. (1969) The diversity of the insect fauna in a hedgerow and neighbouring fields. *Journal of Applied Ecology* **6**, 453–458.

Lisansky, S., Robinson, A. and Coombs, J. (1991) *The UK Green Growers' Guide*. CPL Scientific Press, Newbury.

Loch, J.P.G. (1991) Effect of soil type on pesticide threat to soil/groundwater environment. In: *Chemistry, Agriculture and the Environment* (Ed. M.L. Richardson). Royal Society of Chemistry, Cambridge, pp. 291–307.

MAFF (1998a) *Aerial Applications*. Pesticide usage survey report No 148, Ministry of Agriculture, Fisheries and Food, London.

MAFF (1998b) *Code of Good Agricultural Practice for the Protection of Air*. Revised 1998. Ministry of Agriculture, Fisheries and Foods, London.

MAFF (1998c) *Code of Good Agricultural Practice for the Protection of Soil*. Revised 1998. Ministry of Agriculture, Fisheries and Food, London.

MAFF (1998d) *Code of Good Agricultural Practice for the Protection of Water*. Revised 1998. Ministry of Agriculture, Fisheries and Food, London.

MAFF (1998e) *Code of Practice for the Safe use of Pesticides on Farms and Holdings*. Ministry of Agriculture, Fisheries and Food, London, PB 3528.

MAFF (1999) *Integrated crop management*. Environmental R+D Newsletter, Ministry of Agriculture, Fisheries and Foods, London, No. 5, p. 10–11.

MAFF (2000a) *Farmland bird populations in decline?* Environmental R+D Newsletter, Ministry of Agriculture, Fisheries and Foods, London, No. 6, p. 3.

MAFF (2000b) *Pesticides – how big a risk?* Environmental R+D Newsletter, Ministry of Agriculture, Fisheries and Foods, London, No. 6, p. 11.

MAFF (2000c) *Pesticide poisoning of animals 1999: Investigations of suspected incidents in the United Kingdom*. Ministry of Agriculture, Fisheries and Foods, London, PB 5333.

MAFF (2000d) *Towards Sustainable Agriculture: pilot set of indicators*. Ministry of Agriculture, Fisheries and Food, London, PB 4583.

MAFF (2001) *Birds potentially at risk from seed-pesticides*. Environmental R+D Newsletter, Ministry of Agriculture, Fisheries and Foods, London, No. 7, p. 3.

Makepeace, R. (1996) Sticking up for adjuvants, *Farming and Conservation* April issue, pp. 26–27.

Marshall, T., Holmes, J. and Rose, C. (1999) *Soil Physics*. 3rd Edition, Cambridge University Press, New York.

Mazaud, F. (1997) *Producing enough food is only half the battle*. Group for Assistance on Systems Relating to Grain Harvest. Newsletter No. 20, pp. 4–7.

Megharaj, M., Kookana, R.S. and Naidu, R. (2000) The effect of long-term pesticide contamination on soil biota and their activities. In: *Soil 2000: New Horizons for a New Century. Australian and New Zealand Second Joint Soils Conference Volume 2: Oral Papers* (Eds J.A. Adams and A.K. Metherell). 3–8 December 2000, Lincoln University, New Zealand Society of Soil Science, pp. 205–206.

Merry, R.H., Tiller, K.G. and Alston, A.M. (1986) The effects of soil contamination with copper, lead and arsenic on the growth and composition of plants. *Plant and Soil* **95**, 255–269.

Miller, P.C.H. (1991) Agricultural sprayer design to minimize environmental contamination. In: *Chemistry, Agriculture and the Environment* (Ed. M.L. Richardson). Royal Society of Chemistry, Cambridge, pp. 308–331.

Munz, C. and Bachmann, A. (1993) Documentation of an environmentally-sound soil remediation. In: *Contaminated Soil '93* (Eds F. Arendt, G.J. Annokkee, R. Bosman and W.J. van den Brink). Kluwer Academic Publishers, Netherlands, pp. 1119–1126.

NCC (1990) *Nature Conservation and Agricultural Change*. Report No. 25, Nature Conservancy Council, Peterborough.

Oliver, D.P., Kookana, R.S. and Salama, R. (2000) Effect of land use on sorption behaviour of fenamphos and its metabolites in soils. In: *Soil 2000: New Horizons for a New Century. Australian and New Zealand Second Joint Soils Conference Volume 2: Oral Papers* (Eds J.A. Adams and A.K. Metherell). 3–8 December 2000, Lincoln University, New Zealand Society of Soil Science, pp. 221–222.

Pap, L. and Farkas, R. (1994) Monitoring resistance of insecticides in housefly (*Musca domestica*) populations in Hungary. *Pesticide Science* **40**, 245–258.

Parish, S. (1990) A review of non-chemical weed control techniques. *Biological Agriculture and Horticulture* **7**, 117–137.

Pedersen, H.J., Kudsk, P. and Helweg, A. (1995) Adsorption and ED_{50} values of five soil – applied herbicides. *Pesticide Science* **44**, 131–136.

Pestemer, W. and Krasel, G. (1992) Loss of precipitation from plant and soil by volatilization. In: *Pests and Diseases, Proceedings of the Brighton Crop Protection Conference*, pp. 459–468.

Pimentel, D. (1995) Amounts of pesticides reaching target pests: environmental impacts and ethics. *Journal of Agricultural and Environmental Ethics* **8**, 17–29.

Pollard, E. (1986) Hedges III – The effect of the removal of the bottom flora of a hawthorn hedge on the Carabidae of the hedge bottom, *Journal of Applied Ecology* **5**, 125–139.

Potts, G.R. (1986) *The Partridge: Pesticides, Predation and Conservation.* Collins, London.

Powell, W. (1996) Aphid attack – working on the insides. *Farming and Conservation* April issue, 12–14.

Prosser, P. (2001) *Potential exposure of birds to treated seed.* Final Milestone Report, Project No. PN0907. Central Science Laboratory, York.

PSD (1995) *Consumer Risk Assessment of Insecticide Residues in Carrots.* Pesticides Safety Directorate, York.

Raju, G.S., Khan, S.U. and Millette, J.A. (1993) Pollution potential of selected pesticides in soils. *Chemosphere* **26**, 1429–1442.

Rands, M.R.W. and Sotherton, N.W. (1986) Pesticide use on cereal crops and in the abundance of butterflies on arable farmland in England. *Biology of Conservation* **36**, 71–82.

RCEP (1979) *Agriculture and Pollution.* Seventh Report of the Royal Commission on Environmental Pollution (Cmnd 7644). HMSO, London.

RCEP (1996) *Sustainable Use of Soil.* Nineteenth Report of the Royal Commission on Environmental Pollution (Cm 3165). HMSO, London.

Rush, T. (1994) Herbicide-resistant wild oats discovered. *Arable Farming* September issue, 34–38.

Samuel, A. (1992) UK experiences of 'weed control in the dark'. *New Farmer and Grower* **36**, Autumn issue, 20–21.

Siebers, J., Gottschild, D. and Nolting, H.G. (1994) Pesticides in precipitation in northern Germany. *Chemosphere* **28**, 1559–1570.

Stafford, J.V. and Miller, P.C.H. (1993) Spatially-selective application of herbicide to cereal crops. *Computers and Electronics in Agriculture* **9**, 217–229.

Stehmann, C. and de Waard, M.A. (1995) Relationships between chemical structure and biological activity of triazole fungicides against *Botrytis cinerea. Pesticide Science* **44**, 183–195.

Svec, M., Miklovicova, M., Sykora, M. and Krippel, E. (1995) Fungicide sensitivity of populations of wheat powdery mildew (*Erysiphe graminis* f.sp. *tritici*) in Central Europe in 1993. *Pesticide Science* **43**, 47–52.

Terry, J. (1995) Precision farming – explaining the environmental benefits. *Farming and Conservation* April issue, 8–9.

Turnbull, A. (1996) Chlorinated Pesticides. In: *Issues in Environmental Science and Technology 6: Chlorinated Organic Micropollutants* (Eds R.E. Hester and R.M. Harrison). The Royal Society of Chemistry, Letchworth, pp. 113–135.

van Emden, H.F. (1992) *Pest Control.* Cambridge University Press, Cambridge.

Van Zwieten, L., Ahmad, N., Marolt, R., Ayres, M. and Morris, S. (2001) Determination of DDT and its metabolites in cattle dip soil available in aqueous phase after remediation. *Journal of Environmental Science and Health* **B36**, 501–516.

Warner, M.E. (1995) *An Environmental Risk Index to Evaluate Pesticide Programs in Crop Budgets.* Cornell University, New York.

Wentz, C.A. (1989) *Hazardous Waste Management.* McGraw-Hill, New York.

White, R.E. (1997) *Principles and Practice of Soil Science, the Soil as a Natural Resource.* 3rd Edition. Blackwell Science Ltd, Oxford.

Williams, I.H. (1982) The distribution and decline of British bumble bees. *Journal of Agricultural Research* **21**, 236–245.

Williams, I.H. (1986) Environmental change and the distribution of British bumble bees. *Bee World* **67**, 50–61.

Williams, R.J., Brooke, D.N., Matthiessen, P., Mills, M., Turnbull, A. and Harrison, R.M. (1995) Pesticide transport to surface waters within an agricultural catchment, *Journal of the Institution of Water and Environmental Management* **9**, 72–81.

Wong, S. and Lee, W. (1997) Survey of organochlorine pesticide residues in milk in Hong Kong (1993–1995). *Journal of AOAC International* **80**, 1332–1335.

Worthing, C.R. (1991) *The Pesticide Manual.* 9th Edition. The British Crop Protection Council, Unwin Brothers, London.

Young, R.N., Mohamed, A.M.O. and Warkentin, B.P. (1992) *Principles of Contaminant Transport in Soils.* Elsevier, Amsterdam.

Genetic Modification 8

8.1 INTRODUCTION

The use of biotechnology to genetically modify crops is proving to be highly controversial. Unfortunately, debate has become polarised by 'pro-' and 'anti-' groups (led largely by the biotech industry and environmental campaigners respectively). Ellstrand (2001) states that opinion relating to this argument ranges between 'smug optimism' and 'self-righteous panic'. Environmental campaign groups have adopted emotive phrases such as 'genetic pollution' and 'Frankenfoods' to describe the technology, whilst industry have fought a somewhat unsuccessful campaign to convince the UK public that genetically modified (GM) crops are safe to eat and environmentally benign. In this chapter, we provide some of the evidence on the environmental effects of GM crops. We concentrate on those currently commercially grown and direct the reader to Rissler and Mellon (1996) for discussion of environmental risks associated with GM crops under development but not commercially available (such as those with resistance to viruses). Whether GM crops emerge as 'pollutants' in the same sense as those discussed in Chapters 2–7 of this book remains to be seen.

The use of GM crops world-wide (RAFI, 2000) has increased rapidly, with 1.7 million ha cultivated in 1996, rising to 43 million ha in 2000 (Figure 8.1), grown predominantly in the USA (70%), Argentina (21%), Canada (7%) and China (1%). Four crops currently dominate this market, namely soybeans (58%), maize (12%), cotton (12%) and oilseed rape (7%). GM products already account for 10% of total world seed sales.

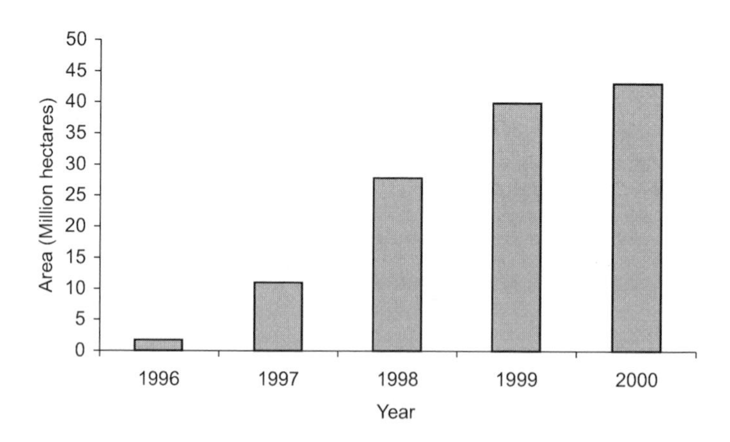

Figure 8.1 Worldwide use of genetically modified crops in hectares from 1996 to 2000 (with kind permission from RAFI, 2000).

The development of crops that are beneficial either in terms of their food value (Box 8.1) or their impact on the environment is clearly of great value. However, opponents of this technology consider that risks to both food safety and the environment outweigh such potential benefits. We do not discuss further the ethical, political and food safety aspects of GM crops, which are beyond the scope of this book. Further information may be obtained from Perry (2001), Reiss and Straughan (1996), Anderson (1999) and The Royal Society's 1998 and 2000 publications as introductions to these topics.

Box 8.1 Golden Rice

Vitamin A deficiency (VAD) is the cause of at least 1 million childhood deaths each year and is considered the single most serious cause of blindness amongst children in the developing countries. UNICEF estimate that some 124 million children around the world are dangerously deficient in vitamin A.

A possible solution to this serious problem is the genetic modification of rice (Potrykus, 2001), a staple food in many developing countries. Rice produces carotenoid compounds that are converted into vitamin A but are present only in the green parts of the plant rather than the grain that is eaten. Professors Potrykus and Beyer have produced a genetically engineered rice variety that contains beta-carotene within the grain itself (hence its 'golden' colour).

The genes that enable the production of beta-carotene within the endosperm come from a daffodil and a bacterium that were inserted using *Agrobacterium* as a vector (Box 8.2). To provide plants that will grow in a particular region's climate, the intention is that golden rice will be crossed with locally grown varieties using conventional plant breeding techniques.

The nutritional benefit of golden rice is coming under intense scrutiny (New Scientist, 2001) and is an illustration of how the debate on these crops is developing. Greenpeace have claimed that carotenoid levels are not sufficiently high in golden rice to provide substantial amounts of dietary vitamin A, stating that 'an adult would have to eat at least 12 times the normal intake of 300 grams to get the recommended daily amount (RDA)' (Greenpeace, 2001). However, this assertion does not consider that insufficient intake rather than a complete absence of RDA causes vitamin A deficiency. Professor Potrykus has pointed out that: 'the amounts required for the prevention of those severe symptoms of vitamin A deficiency are significantly lower than given by RDA-values' and that a diet containing golden rice is beneficial because it provides amounts equivalent to 'the 20–40% range of RDA'.

Genetic information in organisms is held in molecules of DNA. These molecules code a 'blueprint' or specification that allows an organism to function. This blueprint is recorded as genes sequenced along strands of DNA that control biochemical processes. Genetic modification allows the alteration of this blueprint by the insertion of genes (Box 8.1) that cause the expression of a trait that is considered valuable.

Genetic modification is not new; conventional plant breeders have introduced desirable characteristics into our crops for hundreds of years. This is normally done by crossing different varieties of the same species followed by selection, in order

Box 8.2 Gene insertion

This process involves two quite distinct stages. The first is to identify a gene that confers a trait that is considered beneficial, isolate and clone it. The second stage is to transfer this gene from the donor to the target species. Gene insertion may be achieved by a variety of means (Walden and Wingender, 1995

Vectorless gene-transfer

Biolistics DNA is mixed with small particles of tungsten or gold which are bombarded into plant cells.

Electroporation A suspension of plant cells is placed in a container with a solution of DNA containing the gene of interest. A strong electric pulse is then applied which temporarily disrupts cell membranes and allows the DNA to cross into the cells.

Membrane Permeabilisation Plants cells, a solution of DNA and silicon carbide fibres carbide fibres are combined and mixed. The silicon carbide acts to make cell walls permeable, allowing foreign DNA to cross into target cells.

Other vectorless methods include the use of ultrasound or microinjection.

Vector-mediated gene-transfer

Vectors are organisms that are used to transfer a gene from the donor to the target species. Vectors include:

Plasmids Non-chromosomal DNA found in bacteria like *Agrobacterium.*

Viruses e.g. cauliflower mosaic virus.

Transposable Elements Special segments of DNA.

Vector-mediated transfer relies on the use of special enzymes that 'open' specific sections of DNA with a recognised sequence. This process creates sites along the strands of DNA where gene insertion is possible.

to alter the genetic make-up of crops to confer beneficial characteristics. More recently, random DNA mutation has also been used by applying mutagenic chemicals and x-rays followed by selection to improve agronomic characters.

Furthermore, the technique of introgression has been extensively used in cereal and grass breeding programmes and involves hybridising two distantly related species and then backcrossing to the parent until only the desired genes remain in the crop parent.

It should be recognised that introgression occurs naturally between closely related species and is sometimes referred to as 'natural gene flow'.

Genetic modification using biotechnology allows essentially the same process to be achieved; characteristics are conferred on an organism by the selection of a chosen trait. However, characters may be selected from either the same or completely unrelated species or genera and it is this ability which has proven to be so controversial.

The Risk Assessment for Release and Marketing of genetically modified organisms (GMOs) in the European Union includes the assessment of a range of possible impacts (Table 8.1). The environmental safety of GM crops is covered by Directive 90/220/EEC which forms the basis for the regulatory framework adopted in the UK. In the UK, currently no commercial GM crops may be grown, apart from those for assessing environmental impacts. The Advisory Committee on Releases into the Environment (ACRE) regulates the growing of GM crops and is responsible directly to government (http://www.defra.gov.uk/ environment/acre/index.htm for more information). Crops are assessed on biosafety grounds, taking into account the gene donor organism and the effects on the modified organism (Section 8.3).

Table 8.1 Potential hazards associated with GM crops to be considered in Risk Assessment before approval of GMOs (ACRE, 2001).

Expression of toxic or allergenic compounds: the genetic modification may result in the production of substances that are toxic to humans or other species, or allergenic.

Effects on biogeochemistry: the potential to cause changes in nitrogen and carbon recycling that depends on decomposition processes.

Increased persistence in the environment and invasiveness: a genetic modification may confer an ecological fitness advantage to the recipient plant, which potentially allows it to become persistent or invasive. Concerns about 'superweeds' have been raised regarding herbicide tolerance traits; however, traits such as disease, drought or insect resistance are more likely to confer an advantage to a recipient plant, since these pressures control natural plant populations.

Transfer of genetic material: cross-pollination with other crops of the same species or near-relatives can give rise to hybrids which express the traits introduced by the genetic modification. Such gene-transfer may not be a hazard in itself; this would depend on the trait being transferred.

Instability of the genetic modification: plants have the ability to inactivate inserted genetic material, particularly if there is a large number of copies inserted and if the constructs are large. In many cases, this may not pose a risk in itself, as the recipient plant is likely to revert to the wild-type. However, this would become an issue where a genetic modification was made to down-regulate a naturally occurring hazardous trait.

Unintended effects: while it is expected that inserted sequences of nucleic acid are well characterised, the exact positions of the insertion(s) cannot be predicted until more detailed analysis is completed. It is possible that the insertion can influence the expression of adjacent genes and their promoters, leading to unintended genetic modifications. These may not be hazardous, but the transformation event and its progeny would require careful monitoring.

8.2 RISKS AND BENEFITS

In this section, some of the arguments regarding the risks and benefits of commercially grown GM crops are reviewed. It is argued that the potential benefits of these crops are both environmental and agronomic and much discussion of these issues has already occurred (e.g. Barton and Dracup, 2000; Beringer, 2000; Wolfenbarger and Phifer, 2000). Issues relating to two types of GM crop; those that have either herbicide- or insect-resistant properties will be the focus of this section. Both of these crop types potentially offer the opportunity to reduce pesticide use; it has been a long-held goal to develop 'integrated' methods of crop management, minimising the application of agrochemicals and thus reducing the consequent environmental effects.

Herbicide-resistant crops

Crops that are not affected by the application of broad-spectrum herbicides have considerable agronomic advantages because the control of weeds in some crops is both inefficient and expensive. Genetic modification provides a means to confer resistance to broad-spectrum herbicides into crops where current weed control is considered difficult.

Owen (2000) identified the following benefits in this approach as:

- greater flexibility in timing of herbicide application;
- improved weed management;
- use as part of management strategy for herbicide resistant weeds;
- increased opportunities for reduced tillage systems.

Herbicide tolerant crops are not new in themselves; most weed control programmes exploit natural crop resistance to selective herbicides that kill only certain types of plant without harming the crop. Plants may be herbicide-tolerant because they produce (Devine *et al.,* 1993):

- an enzyme which detoxifies the herbicide;
- an altered target enzyme not affected by the herbicide or the plant;
- physical/physiological barriers to uptake of the herbicide.

These characteristics have been identified and through genetic engineering included into GM crop varieties, such as the range of 'Roundup Ready' crops developed by Monsanto (Box 8.2).

Box 8.2 'Roundup Ready' crops

Cotton, soybean, maize, oilseed rape and sugar beet have all been genetically engineered to confer resistance to the broad-spectrum herbicide glyphosate ('Roundup'). The mode of action of this herbicide is to interfere with normal plant metabolism through inhibiting the enzyme 5-enolpyruvyl-3-phosphoshikimic acid synthase (EPSPS). In plants, EPSPS is primarily involved in the biosynthesis of vitamins and aromatic amino acids. Two strategies to produce glyphosate-tolerance are to introduce:

- a glyphosate-tolerant EPSPS gene;
- a gene that allows expression of the enzyme glyphosate oxidoreductase (GOX) that inactivates glyphosate.

Gene insertion has successfully incorporated either tolerant-EPSPS traits or a combination of EPSPS-tolerance and GOX expression (Duke, 1996; Nida *et al.,* 1996; Padgette *et al.,* 1996). The development of herbicide-resistant crops has been rapid; currently there are at least 1000 glyphosate resistant soybean varieties alone with 12–14 million hectares grown in the USA (Lawton, 1999).

In addition to the potential benefits, a number of risks have also been identified (Table 8.2). Environmental benefits are also disputed; Johnson and Hope (2000) consider that 'so far, the industry has produced no convincing evidence that genetically modified herbicide tolerant crops will contribute to minimising the environmental impacts of agriculture'. Evidence available on the environmental and agronomic effects of the commercial use of herbicide-tolerant crops is contradictory with studies both confirming and refuting risks and benefits. Some of this emerging evidence is now reviewed.

One of the potential benefits of herbicide-tolerance is that herbicide application, both in terms of frequency and rate, may be reduced. Glyphosate resistant sugar beet, for example, may need only one or two herbicide applications compared to up to eight applications of a cocktail of conventional herbicides (Dewar *et al.,* 2000a). In the case of soybeans, however, there is evidence that herbicide-resistant varieties may require more herbicides than conventional ones. Wolfenbarger and Phifer (2000) conclude that although GM crops have resulted in an overall reduction in the use of chemicals, this is not evident in soybean.

A review of the current use of transgenic herbicide-tolerant soybean in the USA concludes that Monsanto's initial claim that growers could use just one application of glyphosate to provide weed control had not been achieved (Owen, 2000). In a detailed study of herbicide use, Benbrook (2001) demonstrated that herbicide use in transgenic herbicide-tolerant soybean exceeded that of conventional crops overall (Table 8.3) and reports a price reduction in the use of herbicides which has encouraged farmers to increase applications. The increased reliance on herbicides can cause shifts in weed assemblages (Ghersa *et al.,* 1994) that make control more difficult; hence, the recurrent and dependent use of glyphosate may, over time, result in weed control problems.

Table 8.2 Potential benefits and risks of transgenic herbicide-resistant crops (Kuiper *et al.*, 2000). With permission from Elsevier Science.

Potential Benefits	Potential risks
Simpler weed management based on fewer herbicides	Greater reliance on herbicides for weed control
Decrease in herbicide use	Increase in herbicide use
Less contamination of the ecosystem	More contamination of the water, soil and air and shift in exposure patterns
Use of environmentally benign herbicides	Development of resistance in weed species by introgression of transgenes
Reduction of the need for mechanical soil treatment	Shift in population of weeds towards more tolerant species
Less crop injury	Increase in volunteer problems
Improved weed control	Negative effects of on non-target species

Table 8.3 Herbicide applied per hectare in Roundup Ready and conventional soybean varieties for all surveyed states and five example states surveyed by USDA in 1998 (Benbrook, 2001). With permission from AgBioTechInfoNet.

State	Total herbicides (kgha^{-1})		Ratio
	RR soybean	Conventional	RR:conventional
ALL SURVEYED STATES	2.68	2.38	1.13
Arkansas	3.30	2.02	1.63
South Dakota	3.12	2.11	1.48
Kentucky	2.46	2.40	1.03
Nebraska	2.73	3.19	0.86
Michigan	2.27	3.23	0.70

UK conservation organisations such as the Royal Society for the Protection of Birds and English Nature have raised concerns about the potential widespread use of herbicide-tolerant crops as it could lead to a 'cleaner' weed-free crop which might have a negative impact on biodiversity. Conversely, herbicide tolerant crops may allow a shift towards minimum tillage techniques that would benefit the environment (Wolfenbarger and Phifer, 2000) because they:

- may decrease soil erosion;
- can reduce water loss;
- may increase soil organic matter;
- allow herbicide usage to be reduced.

The transfer of genes to weedy relatives, resistant volunteers acting as weeds and transfer of tolerance between varieties resulting in multiple-resistance have all been identified as potential environmental risks (Hall *et al.,* 2000). The risk of genes conferring herbicide resistance to weed species due to out-crossing (Box 8.3) and the consequent emergence of 'superweeds' is one of the key environmental arguments against the use of genetic engineering.

Box 8.3 Outcrossing

The risk of out-crossing of resistant genes varies according to the way in which a crop is pollinated. A crop may be self-or cross-pollinated (or both) by insect or wind. Crops such as oilseed rape are insect pollinated and partially cross-breeding plants.

Outcrossing may occur:

- between varieties of the same species. For example, *B. napus* has an interplant outcrossing rate of 21.8% (Rakow and Woods, 1987);
- by the transfer of genetic information from crop $\rightarrow$ weed in systems such as cultivated radish-wild radish (Klinger *et al.,* 1991) and sorghum-johnsongrass (Arriola and Ellstrand, 1996; Ellstrand *et al.,* 1999).

The likelihood of outcrossing reduces with distance (Stringham and Downey, 1978); in field trials, Downey (1999) showed that *B. napus* pollen transfer results in under 1% out-crossing within 100m from source although rape pollen can travel 1.5 to 2.5 km from commercial fields (Timmons *et al.,* 1996). In trials of transgenic cotton, outcrossing averaged from 0.15% of progeny at 1 m to below 0.08% at 4 m from the test plots (Llewellyn and Fitt, 1996). This risk of outcrossing has led to the use of buffer zones in field trials of transgenic crops during assessments of their environmental effects (e.g. Staniland *et al.,* 2000).

It should be noted that crop-to-weed or variety-to-variety gene flow (introgression) is not found solely in transgenic plants; this process has occurred in conventional agricultural systems and has caused 'superweed' problems such as those faced in Europe with the sea beet-sugar beet hybrid (van Dijk and Desplanque, 1999). It is the risk of genetically engineered traits such as herbicide-tolerance escaping into wild populations that is specific to transgenic plants. Consequently, some crops such as sorghum that hybridise fairly easily may be too risky to genetically engineer because of likely hybridisation with wild relatives like johnsongrass, a pernicious weed (Kuiper *et al.,* 2000). The risk of crop-to-weed gene flow may be assessed using a stepwise approach. Figure 8.2 shows steps identified by Rissler and Mellon (1996) that are necessary when assessing the potential of transgene flow.

Tier 1: Gene flow

Do viable, fertile hybrids form between the crop and wild/weedy relatives ?

a. Is the crop sexual ?

Yes/Insufficient information	No
Go to b	Lower risk – end of analysis

b. Does the crop have sexually compatible wild/weedy relatives ?

Yes/Insufficient information	No
Go to c	Lower risk – end of analysis

c. Do the crop-relatives breeding systems permit gene flow in and out?

Yes/Insufficient information	No
Go to d	Lower risk – end of analysis

d. Does the flowering phenology of the crop and wild/weedy relatives overlap?

Yes/Insufficient information	No
Go to c	Lower risk – end of analysis

e. Do crop and wild/weedy relatives share the same means of pollination ?

Yes/Insufficient information	No
Go to f	Lower risk – end of analysis

f. Do crop and wild/weedy relatives naturally cross-pollinate, fertilise, and set viable, fertile seeds under field conditions ?

Yes/Insufficient information	No
Go to tier 2	Lower risk – end of analysis

Tier 2: Ecological performance of transgenic wild/weedy plants. Do transgenic wild/weedy populations outperform the non-transgenic wild/weedy plants in population replacement experiments?

Yes	No
High risk: rethink commercialisation or go to tier 3	Lower risk – end of analysis

Tier 3: Weediness of transgenic wild/weedy plants. Is weediness increased in transgenic wild/weedy plants exhibiting enhanced ecological performance ?

Yes	No
High risk: rethink commercalisation	Lower risk – end of analysis

Figure 8.2 Experimental assessment of the potential for transgene flow to produce weeds in non-crop populations. Three tiers provide a structured pathway to analyse the potential risk of gene flow (Rissler and Mellon, 1996).

One example of gene flow was reported by Hall *et al.* (2000). In 1997, a grower in Alberta, Canada planted glufosinate- and imidazolinone-resistant *B. napus* in two sections of the same field (Field 1). A glyphosate-resistant variety was planted in an adjacent field (Field 2) separated by a 22 m road. In 1998, part of Field 2 was planted with imidazolinone-resistant *B. napus* whilst Field 1 was fallowed and glyphosate treatments applied for weed control. *B. napus* volunteer plants, unaffected by glyphosate, were reported in Field 1 without the resistant variety being sown. A detailed study concluded that pollen flow was responsible for the occurrence of multiple-resistance; triple-resistant mutations did occur and a single triple-resistant individual was located over 550 m from the pollen source after about 18 months. It is important to consider whether the presence of this effect is of any major environmental concern; the workers in this study conclude that integrated management, such as using crop and herbicide rotations could be used to reduce the incidence of volunteers with multiple resistance characteristics.

In fact, the use of transgenic varieties may make it easier to control volunteers; Dewar *et al.* (2000b) showed that glyphosate applied early, late, or twice to genetically modified herbicide-tolerant sugar beet gave excellent control of potato volunteers. A significant reduction in the number and size of daughter tubers produced helped to prevent volunteers in the next crop rotation. The approach had the environmental benefit of reducing the number of chemical applications from between two-five down to one-two, in addition to suppression of potato cyst nematode populations. The long-term effects of using herbicide-tolerant transgenic plants remain largely unknown; there is evidence of agronomic and environmental benefit as well as potential risk, and increased complexity in crop management.

The potential increase in yield of the world's crops has often been cited as one of the key benefits of transgenics. With an increasing world population it is coherently argued that technology which increases yield has enormous value, firstly to ensure that global food production needs are met and secondly that it is done without the need to cultivate an increasingly large land area which in itself would cause environmental degradation. The effect that herbicide-tolerant crops have on yield vary between crops; the USDA (United States Department of Agriculture) report increase in both yield and profits for herbicide-tolerant cotton and a small increase in yield for soybean for data collected in 1997 (ERS/USDA, 2001). Some studies indicate that yield losses may be occurring in certain crops which may be caused by two phenomena (IANR, 2000):

Drag Lower yields resulting directly from gene insertion.

Lag Loss in yield due to the crop variety used for inclusion of trait.

In the case of commercially grown crops such as soybean, the yield benefits are often ambiguous. In a detailed study, Benbrook (2001) concluded that Roundup Ready soybean yields were 5 to 10% less than conventional crops grown under identical conditions (Table 8.4). This yield drag was attributed to the impairment of root development, nodulation and nitrogen fixation due to the bacterial symbiont (*Bradyrhizobium japonicum*) responsible for nitrogen fixation being sensitive to Roundup.

Table 8.4 Yields (bushels per acre) and yield drag (as percentage, negative values indicating relatively lower, and positive relatively higher yields respectively in Roundup Ready crops) for conventional and Roundup Ready soybean in variety trials tested at 9 locations in Illinois (Benbrook, 2001). With permission from AgBioTechInfoNet.

Trial	Conventional	Roundup Ready	Drag (%)
Perry	55.5	54.0	–2.7
Dwight	68.2	67.2	–1.5
New Berlin	68.7	66.7	–3.0
Monmouth	59.2	61.6	+3.3
Erie	69.0	68.6	–0.5
Goodfield	64.1	60.6	–5.4
DeKalb	72.5	71.0	–2.1
Dixon Springs	65.4	58.1	–11.1
Urbana	62.8	64.4	+2.6

Insect-resistant crops

The genetic modification of crops to confer resistance against insect attack is well advanced and varieties for a number of crops are commercially available, primarily in the USA. The expression of an insect toxin from *Bacillus thuringiensis* (Box 8.4) has provided the first insect-resistant transgenic crops. These varieties are not intended to increase yield, rather they prevent yield loss when pest species are sufficiently abundant. If resistance can be conferred to crops then there are potential benefits in the reduced use of insecticides in the environment. The use of insect-resistant Bt crops has rapidly increased, particularly in the USA (Table 8.5); maize, cotton and potatoes covered about 16 million acres in 1998 (Gianessi and Carpenter, 1999).

Box 8.4 *Bacillus thuringiensis*

Bacillus thuringiensis (Bt) is a soil-living organism that may be used conventionally to control a range of insect pests (Chapter 7). Bt needs to be ingested by the target insect; it contains a crystalline toxin (encoded by Cry genes) that paralyses the insect's digestive tract, causing death.

Crops have been genetically engineered to contain the toxic component of Bt. Because the genetic code of plants and bacteria are dissimilar, genes are synthesised that make the bacterial amino acid sequence and the target plant DNA compatible. The Bt toxin is not soluble in plant cells because of their relatively low pH. This is overcome by using a shortened gene sequence that produces toxin molecules that are soluble in plant cells; hence the toxin is present in the green plant tissues itself.

From 1996 crops including maize, potato, cotton and soybean have been engineered to incorporate the Bt toxin and are available commercially in the US and elsewhere. All express part of the Cry toxin in their tissues. The gene for this toxin has been introduced in an effort to confer protection against insect attack without the need for conventional insecticides.

Table 8.5 US crop acreage planted with Bt crops in 1998 (Gianessi and Carpenter, 1999).

Crop	Million acres	% of crop
Field Corn	14.40	18
Cotton	2.30	17
Potatoes	0.05	4

The environmental impact of conventional insecticide use, particularly those with broad-spectrum activity, is well documented and covered in detail in Chapter 7. Genetic engineering, such as the inclusion of the Bt toxin, provides an opportunity to confer insect-resistant properties in the plants themselves, which could lead to a reduction in conventional pesticide use. Whether current commercially available transgenic crops have delivered this environmental benefit is unclear with much contradictory evidence available. These contradictory conclusions are illustrated well in cotton and corn.

The US Environmental Protection Agency (EPA and USDA, 1999) state that 'Bt toxins expressed in transgenic plants can and have reduced the use of more hazardous insecticides such as organophosphates and synthetic pyrethroids'. They further claim that 'since the commercialisation of Bt cotton in 1996, industry information has shown that cotton insecticide use has been reduced by approximately 1 million gallons of formulated product per year'. However, in a report compiled for the WWF (2000) authors claim that 'From 1996 to 1998 the acreage of Bt-cotton steadily increased reaching 17% of the total cotton acreage in the U.S. in 1998, while insecticide use per acre remained more or less at the same level'.

Obrycki and Losey (2001) questions the widespread use of Bt corn, stating that it has not reduced pesticide use or increased yields. Burkness *et al.* (2001), however, concluded that Bt sweet corn hybrids without the use of conventional insecticides could provide excellent control against two lepidopteran pests *Ostrinia nubilalis* and *Helicoverpa zea*.

A major concern regarding the use of these transgenic crops is that the technology may promote resistance in the pest species themselves, ultimately making control more difficult leading to increased use of conventional insecticides. The process by which resistance emerges in insect populations is well understood and described in Chapter 7. Additionally, biopesticides such as Bt are used in organic farming systems; there is concern that this control method may be lost to such growers if insecticide-resistance results from the inclusion of Bt toxin expression in transgenic crops.

Tabashnik (1994) has demonstrated that resistant traits to Bt in insect populations may emerge and consequently there is concern that the wide use of transgenic Bt crops may accelerate the expression of resistance. The management of insect resistance is not a problem exclusive to transgenic crops and it is possible that their use may increase our ability to reduce the emergence of resistance by integrated management methods. The possible emergence of insect-resistance has been taken very seriously because it would mean that the transgenic crops would become ineffective. Consequently, strategies adopting the use of refuges to manage resistance have been devised. Refuges are non-transgenic crop areas where insect populations without resistant traits may survive. This provides an insect population which suppresses the expression of resistant traits within the pest population as a whole. An example of resistance management using refuges is that agreed by the US EPA and developed by members of the Agricultural Biotechnology Stewardship Technical Committee (ABSTC, 2001). The plan incorporates a 20% refuge requirement in the major corn-growing regions and a 50% refuge requirement in areas of overlapping corn and cotton production.

Another technique to manage this problem is 'stacking' or 'pyramiding', where two contrasting insecticidal characteristics are included in the genetically engineered plant (Roush, 1998). This reduces the likelihood of resistance emerging within a pest population, thus ensuring the long-term efficacy of the transgenic plant.

The use of genetically engineered crops has been criticised because they potentially threaten non-target organisms (Butler and Reichhardt, 1999) which may either be beneficial to agriculture (e.g. natural enemies) or simply of conservation interest (e.g. butterflies). Much work has been done in the laboratory and at the field scale to investigate these potential effects on a wide range of non-target species (Hilbeck *et al.*, 2000; Schuler, 2000), but it is work on the monarch butterfly that has stimulated so much media interest (Box 8.5).

Box 8.5 The monarch butterfly

Monarch butterflies (*Danaus plexippus*) migrate each year from Mexico to the USA and Canada (navigating a flight of 2000 miles), laying its eggs on the underside of leaves of the milkweed plant. Studies have been carried out to investigate the potential impacts of Bt corn on these insects (e.g. Palevitz, 1999; Losey *et al.*, 1999; Jesse and Obrycki, 2000). These studies were done due to the potential exposure of monarch butterflies to corn pollen expressing Bt toxin in areas where milkweed plants are common.

The study by Jesse and Obrycki (2000) investigated Bt modified corn, which was engineered to control a pest called the European corn borer. They investigated whether Bt corn pollen landing on the larval host plant of milkweed might pose a risk. The experiment was conducted by removing leaf disks from milkweed plants that had been grown adjacent to Bt and non-Bt corn and then allowing newly hatched larvae to feed on them in the laboratory. The workers showed that larvae exposed for 48 hours to Bt corn that had accumulated on milkweed in the field showed 20% mortality compared to 0% mortality for control (not-Bt corn) plants. Overall, the study concluded that significant mortality of monarch larvae occurred from natural pollen deposition from Bt modified crops and further study was needed to assess such effects in the field.

However, the interpretation of these data has been questioned by some workers. In a critique of this study Sears and Shelton (2000), conclude that the study 'does little to help understand potential risks of deploying Bt plants in the field'. The findings from the study were questioned primarily because:

- toxicity results were not consistent with other studies;
- small sample size (35 larvae were exposed to Bt corn) making results difficult to interpret;

- concentrations of pollen grains that cause high mortality were unlikely to be evident under field conditions;
- larvae were 'forced' to eat contaminated leaf tissue and had no choice of diet.

In a series of papers published in the Proceedings of the National Academy of Sciences of the USA, Hellmich *et al.* (2001), Sears *et al.* (2001) and Stanley-Horn *et al.* (2001) assessed the risk of Bt to monarchs in the field. The overall conclusion of the study indicated that the impact of Bt corn pollen from commercially available hybrids was negligible. The key reasons for this were:

- low Bt expression corn pollen in most commercial GM corn resulting in no acute toxic effects observable in laboratory or field trials;
- limited overlap in time between pollen shed and larval activity;
- only a proportion of the monarch population utilising milkweed near corn fields;
- 19% adoption of Bt corn in North America.

These studies illustrate the need for work which determine the likelihood and risk of exposure of non-target species to such toxins. Whilst the study by Jesse and Obrycki (2000) demonstrates a potential mortality pathway, it is the translation of such an effect into observable field mortalities that must be established before risks can be established. Studies on another butterfly, the black swallowtail *Papilio poyxenes* indicated that mortality did not occur in laboratory trials even when insects were exposed to very high Bt corn pollen doses (Wraight *et al.*, 2000).

A study on generalist predator populations present in transgenic Bt potatoes compared to conventional varieties concluded that there were no deleterious effects (Riddick *et al.*, 2000). There is a concern that the insecticidal effects of Bt crops may cause problems by their persistence, although studies have shown rapid degradation of Bt toxin activity in neutral soils (Sims and Holden, 1996). Studies also suggest that Bt toxin may enter the soil ecosystem by exuding from Bt corn roots (Saxena *et al.*, 1999) and that Bt toxin uptake from soil by plants is possible (Saxena and Stotzky, 2001) although the implications of these observations in the field remain to be quantified fully.

8.3 ASSESSING ENVIRONMENTAL RISK

The examples above illustrate the range of concerns raised regarding the use of transgenic crops. In order to minimise potential problems when they are released into the environment, a process of risk assessment is undertaken. This process should provide an opportunity to determine environmental effects in order to determine their suitability for commercial use. The process is complex and an illustration of factors that need to be considered is given in Figure 8.2.

Studies to assess risk include a combination of:

- bioassays – small scale studies conducted in a laboratory usually conducted under 'worst case' scenarios;
- laboratory experiments that mimic field conditions and ecosystems;
- small scale field trials;
- large scale field trials.

No genetically engineered crops have completed all the UK regulatory requirements allowing them to be grown for commercial purposes. Herbicide tolerant maize, oilseed rape and fodder beet are those closest to commercialisation. These crops are now being assessed in large-scale experiments to determine their overall effect on biodiversity. Biodiversity may be defined as 'the genetic, taxonomic and ecosystem variety in living organisms of a given area, environment, ecosystem or the whole planet' (McAllister, 1991). One of the key concerns of environmental groups is that transgenic crops may disrupt agricultural ecosystems thus causing detrimental effects on biodiversity.

As part of the registration of such crops, the EU requires that an assessment of all direct, indirect, immediate and delayed effects of the GM crop to the environment is carried out before commercial growing. The UK government have adopted a 'precautionary approach' to GM technology and have commissioned a number of studies to gauge the environmental risks of using this technology (Firbank *et al.*, 1999). During 2001, genetically modified rape, maize and beet will be grown across the UK as part of the FSEs (current information regarding these trials may be found at http://www.defra.gov.uk/environment/fse/index.htm). These FSEs were agreed between the farming and biotechnology industry body, SCIMAC (the Supply Chain Initiative on Modified Agricultural Crops) and trials will continue until the end of 2003. The farm-scale evaluations are concerned with determining the effects of transgenic crops on biodiversity. The FSE programme is designed to 'examine whether there are any differences in the diversity and abundance of farmland wildlife associated with the management of GM herbicide tolerant crops as compared with equivalent non-GM crops'. It is imperative that field-scale trials are done because small-scale evaluation in the laboratory or semi-field scale does not incorporate the complexity of 'real' ecosystems. The study started in 1999 and field trials are being conducted on the ecological effects and a wide range of factors are being measured (Table 8.6) as environmental indicators of change.

The FSE is designed to detect differences in biodiversity between GM and conventional crops. Positive, negative, or no effect may be detected. The trials are made up of about 25 fields per crop (fodder maize, spring and winter oilseed rape, and sugar beet) using a 'split field' design with one half of the field having a GM crop and the other half a conventional variety. It is intended that the findings from these studies will be published rapidly so informed decision-making regarding the risks of growing transgenic crops in the UK may be carried out.

Table 8.6 Environmental indicators measured during UK Farm Scale Evaluations (DEFRA, 2001).

Soil seed bank.
Arable plant diversity, biomass and estimated seed return.
Field margin and boundary vegetation, noting species in flower and signs of unintentional spray drift of weed-killer during its application to the field.
Gastropods (slugs and snails): abundance, activity and diversity measures.
Arthropods (especially insects and spiders) on vegetation, concentrating on plant bugs (Heteroptera), spring tails (Collembola), and the caterpillars of butterflies, moths, (Lepidoptera) and sawflies (Hymenoptera); diversity and biomass measures.
Ground (Carabid) beetles and other ground dwellings arthropods; abundance and diversity measures.
Bees and butterflies; observational studies.
Birds and mammals; observational studies.

8.4 THE FUTURE FOR GMOs

This chapter has been included in order to highlight how advances in technology create new environmental challenges and potential pollution problems. Transgenic crops are of potentially enormous value agronomically, but there is uncertainty regarding the environmental consequences of their use. The assessment of risk is incomplete and much more evidence is needed before the impact of genetically modified crops is fully understood. This is complicated further by the need for a comparison with conventional practices that may themselves lead to environmental harm. It is only as studies are completed and published that an informed assessment of the possible environmental implications can be made.

It is also clear that the GM crops currently under scrutiny are just the first wave of modified varieties. Already, efficiency of GM technology has advanced and greater attention been paid to both the expression of the modified character and its potential for 'escape' into the environment. Clearly ecological and environmental evaluation of impacts of GM crops needs to be accelerated to keep pace with these biotechnological advances.

REFERENCES

ABSTC (2001) *Bt Corn Insect Resistance Management Survey – 2000 growing season*. Agricultural Biotechnology Stewardship Technical Committee. URL:http://www.biotech-info.net/bt_corn_survey_2000.pdf.

ACRE (2001) *Risk Assessment for Release and Marketing of GMOs in the European Union*. URL: http://www.defra.gov.uk/environment/acre/background/risk/index.htm

Anderson, L. (1999) *Genetic Engineering, Food and our Environment*. Green Books, Dartington.

Arriola, P.E. and Ellstrand, N.C. (1996) Crop-to-weed gene flow in the genus Sorghum (Poaceae): spontaneous interspecific hybridization between johnsongrass. *Sorghum halapense*, and crop sorghum, *S. bicolor*. *American Journal of Botany* **83**, 1153–1160.

Barton, J.E. and Dracup, M. (2000) Genetically modified crops and the environment. *Agronomy Journal* **92**, 797–803.

Benbrook, C.M. (2001) Troubled times amid commercial success for Roundup Ready Soybeans. AgBioTechInfoNet Technical Paper Number 4. URL: http://www.biotech-info.net/troubledtimes.html

Beringer, J.E. (2000) Releasing genetically modified organisms: will any harm outweigh any advantage? *Journal of Applied Ecology* **37**, 207–214.

Burkness, E.C., Hutchison, W.D., Bolin, P.C., Bartels, D.W., Warnock, D.F. and Davis, D.W. (2001) Field efficacy of sweet corn hybrids expressing a *Bacillus thuringiensis* toxin for management of *Ostrinia nubilalis* (Lepidoptera: Crambidae) and *Helicoverpa zea* (Lepidoptera: Noctuidae). *Journal of Economic Entomology* **94**, 197–203.

Butler, D. and Reichardt, T. (1999) Assessing the threat to biodiversity on the farm. *Nature* **398**, 654–656.

DEFRA (2001) *Genetically modified crop Farm-Scale Evaluations: Some questions answered about GM crops and the Farm Scale Evaluations*. Department of Environment, Food and Rural Affairs. URL: http://www.defra.gov.uk/environment/fse/raq/index.htm

Devine, M., Duke, S.O. and Fedtke, C. (1993) *Physiology of Herbicide Action*. Prentice Hall, Englewood Cliffs, NJ, pp. 251–294.

Dewar, A.M., Haycock, L.A., May, M.J., Beane, J. and Perry, R.N. (2000b) Glyphosate applied to genetically modified herbicide-tolerant sugar beet and 'volunteer' potatoes reduces populations of potato cyst nematodes and the number and size of daughter tubers. *Annals of Applied Biology* **136**, 179–187.

Dewar, A.M., May, M.J. and Pidgeon, J.D. (2000a) GM sugar beet – the present situation. *British Sugar Beet Review* **68**, 22–27.

Downey, R.K. (1999) Risk assessment of out-crossing of transgenic *Brassica*, with focus on *B. rapa* and *B. napus*. *Proceedings of the 10th International Rapeseed Congress,* Canberra, Australia.

Duke, S.O. (1996) *Herbicide Resistant Crops*. CRC Press, New York.

Ellstrand, N.C. (2001) When transgenes wander, should we worry? *Plant Physiology* **125**, 1543–1545.

Ellstrand, N.C., Prentice, H.C. and Hancock, J.F. (1999) Gene flow and introgression from domesticated plants into their wild relatives. *Annual Review of Ecological Systems* **24**, 217–242.

EPA and USDA (1999) Position Paper on Insect Resistance Management in Bt Crops 5/27/99 (minor revisions 7/12/99) URL: http://www.epa.gov/oppbppd1/biopesticides/otherdocs/bt_position_paper_618.htm

ERS and USDA (2001) Emphasis – Harmony between agriculture and the environment: genetically modified crops. URL: http://www.ers.usda.gov/Emphases/Harmony/ issues/genengcrops/terms.htm

Firbank, L.G., Dewar, A.M., Hill, M.O., May, M.J., Perry, J.N., Rothery, P., Squire, G.R. and Woiwod, I.P. (1999) Farm-scale evaluation of GM crops explained. *Nature* **399**, 727–728.

Ghersa, C.M., Roush, M.L., Radosevich, S.R. and Cordray, S.M. (1994) Coevolution of agroecosystems and weed management. *Bioscience* **44**, 85–94.

Gianessi, L.P. and Carpenter, J.E. (1999) Agricultural Biotechnology: Insect Control Benefits, NCFAP Report, July 12, 1999 URL: http://www.bio.org/foodandag/ncfap/ag_bio.htm.

Greenpeace (2001) Genetically engineered 'golden rice' is fool's gold. URL: http://www.greenpeace.org/pressreleases/geneng/2001feb9.html

Hall, L., Topinka, K., Huffman, J., Davis, L. and Good, A. (2000) Pollen flow between herbicide-resistant *Brassica napus* is the cause of multiple-resistant *B. napus* volunteers. *Weed Science* **48**, 688–694.

Hellmich, R.L. Siegfried, B.D. Sears, M.K., Stanley-Horn, D.E., Daniels, S, M.J., Mattila, H.R., Spencer, T., Bidne, K.G. and Lewis, L.C. (2001) Monarch larvae sensitivity to *Bacillus thuringiensis*-purified proteins and pollen. *Proceedings of the National Academy of Sciences of the USA* **98**, 11925–11930.

Hilbeck, A., Meier, M.S. and Raps, A. (2000) *Review on non-target organisms and Bt plants*. Report to Greenpeace International by Ecological Technology Assessment and Environment Consulting. URL: http://www.greenpeace.org/~geneng/reports/gmo/ecostratbt.pdf

IANR (2000) *Research shows Roundup Ready Soybeans Yield Less*. University of Nebraska, Institute of Agriculture and Natural Resources publication, July 2000. URL: http://www.biotech-info.net/Roundup_soybeans_yield_less.html

Jesse, L.C.H. and Obrycki, J.J. (2000) Field deposition of Bt transgenic corn pollen: lethal effects on the monarch butterfly. *Oecologia* **125**, 241–248.

Johnson, B. and Hope, A. (2000) GM crops and equivocal environmental benefits. *Nature Biotechnology* **18**, 242.

Klinger, T., Elam, D.R. and Ellstrand, N.C. (1991) Radish as a model system for the study of engineered gene escape rates via crop-weed mating. *Conservation Biology* **5**, 531–535.

Kuiper, H.A., Kleter, G.A. and Noordam, M.Y. (2000) Risks of the release of transgenic herbicide-resistant plants with respect to humans, animals, and the environment. *Crop Protection* **19**, 773–778.

Lawton, K. (1999) Roundup of a market. *Farm Industry News*, February, 4–8.

Llewellyn, D. and Fitt, G. (1996) Pollen dispersal from two field trials of transgenic cotton in the Namoi Valley, Australia. *Molecular Breeding* **2**, 157–166.

Losey, J.E., Rayor, L.S. and Carter, M.E. (1999) Transgenic pollen harms monarch larvae. *Nature* **399**, 214.

McAllister, D.E. (1991) What is biodiversity? *Canadian Biodiversity* **1**, 4–6.

New Scientist (2001) Protests take the shine off golden rice. *New Scientist* **169** Issue 2284, 15.

Nida, D.L., Kolacz, K.H., Buehler, R.E., Deaton, W.R., Schuler, W.R., Armstrong, T.A., Taylor, M.L., Ebert, C.C. and Rogan, G.J. (1996) Glyphosate-tolerant cotton: Genetic characterization and protein expression. *Journal of Agricultural and Food Chemistry* **44**, 1960–1966.

Obrycki, J.J., Losey, J.E., Taylor, O.R. and Jesse, L.C.H. (2001) Transgenic Insecticidal Corn: Beyond Insecticidal Toxicity to Ecological Complexity *Bioscience* **51**, 353–362.

Owen, M.D.K. (2000) Current use of transgenic herbicide-resistant soybean and corn in the USA. *Crop Protection* **19**, 765–771.

Padgette, S.R., Re, D.B., Barry, G.F., Eichholtz, D.E., Delannay, X., Fuchs, R.L., Kishore, G.M. and Fraley, R.T. (1996) New weed control opportunities: Development of soybeans with a Roundup Ready gene. In: *Herbicide-Resistant Crops: Agricultural, Environmental, Economic, Regulatory, and Technical Aspects* (Ed. S.O. Duke). CRC Press Inc., Boca Raton, Florida, and London, England, pp. 53-84.

Palevitz, B.A. (1999) Bt or not Bt… Transgenic corn vs. monarch butterflies. *Scientist* **13**, 1.

Perry, J.N. (2001) Genetically Modified Crops. In: *Genetic Engineering, Volume XV in the 'Christ and the Cosmos' series* (Ed. B. Beamond). *Proceedings of the 15th Christ and the Cosmos Conference*, London Colney, Herts.

Potyrkus, I. (2001) Golden rice and beyond. *Plant Physiology* **125**, 1157–1161.

RAFI (2000) Speed Bump or Blow-Out for GM seeds ? URL: http://www. rafi.org/web/docus/pdfs/Gmseed2000.pdf

Rakow, G. and Woods, D. (1987) Out-crossing in rape and mustard under Saskatchewan prairie conditions. *Canadian Journal of Plant Science* **67**, 147–151.

Reiss, M.J. and Straughan, R. (1996) *Improving Nature? The Science and Ethics of Genetic Engineering*. Cambridge University Press.

Riddick, E.W., Dively, G. and Barbosa, P. (2000) Season-long abundance of generalist predators in transgenic versus non-transgenic potato fields. *Journal of Entomological Science* **35**, 349–359.

Rissler, J. and Mellon, M. (1996) *The Ecological Risks of Engineered Crops.* The MIT Press, Cambridge, Massachusetts.

Roush, R.T. (1998) Two-toxin strategies for management of insecticidal transgenic crops: Can pyramiding succeed where pesticide mixtures have not? *Philosophical Transactions of the Royal Society of London* **B 353**, 1777–1786.

Royal Society (1998) Genetically Modified Plants for Food Use. Documents 1/98 and 2/98.

Royal Society (2000) Transgenic Plants and World Agriculture. Document 08/00.

Saxena, D., Flores, S. and Stotzky, G. (1999) Transgenic plants – Insecticidal toxin in root exudates from Bt corn. *Nature* **402**, 480.

Saxena, D. and Stotzky, G. (2001) Bt toxin uptake from soil by plants. *Nature Biotechnology* **19**, 199.

Schuler, T.H. (2000) The impact of insect resistant GM crops on populations of natural enemies. *Antenna* **24**, 59–65.

Sears, M. and Shelton, A. (2000) *Questionable conclusions from the latest monarch study.* URL: http://www.biotech-info.net/questionable_conclusions.html

Sears, M.K., Hellmich, R.L., Stanley-Horn, D.E., Oberhauser, K.S., Pleasants, J.M., Matilla, H.R., Siegfried, B.D. and Dively, G.P. (2001) Impact of Bt corn pollen on monarch butterfly populations: A risk assessment. *Proceedings of the National Academy of Sciences of the USA* **98**, 11937–11942.

Sims, S.R. and Holden, L.R. (1996) Insect bioassay for determining soil degradation of *Bacillus thuringiensis* subsp *kurstaki* CryIA(b) protein in corn tissue. *Environmental Entomology* **25**, 659–664.

Staniland, B.K., McVetty, P.B.E., Friesen, L.F., Yarrow, S., Freyssinet, G. and Freyssinet, M. (2000) Effectiveness of border areas in confining the spread of transgenic *Brassica napus* pollen. *Canadian Journal of Plant Science* **80**, 521–526.

Stanley-Horn, D.E., Dively, G.P. Hellmich, R.L Matilla, H.R., Sears, M.K., Rose, R., Jesse, L.C.H., Losey, J.E., Obrycki, J.J. and Lewis, L. (2001) Assessing the impact of Cry 1 Ab-expressing corn pollen on monarch butterfly larvae in field studies. *Proceedings of the National Academy of Sciences of the USA* **98**, 11931–11936.

Stringham, G.R. and Downey, R.K. (1978) Effectiveness of isolation distance in turnip rape. *Canadian Journal of Plant Science* **58**, 427–434.

Tabashnik, B.E. (1994) Evolution of resistance to *Bacillus thuringiensis. Annual Review of Entomology* **39**, 47–79.

Timmons, A.M., Charters, Y.M., Crawford, J.W., Burn, D., Scott, S.E., Dubbels, S.J., Wilson, N.J., Robertson, A., O'Brien, E.T., Squire, G.R. and Wilkinson, M.J. (1996) Risks from transgenic crops. *Nature* **380**, 487.

vanDijk, H. and Desplanque, B. (1999) *European Beta: crops and their wild and weedy relatives.* 7th International Symposium of the International-Organization-of-Plant-Biosystematics, August 10–15, 1998. Plant Evolution in Man-Made Habitats, 257–270.

Walden, R. and Wingender, R. (1995) Gene-transfer and plant--regeneration techniques. *Trends in Biotechnology* **13**, 324–331.

Wolfenbarger, L. L. and Phifer, P.R. (2000) The ecological risks and benefits of genetically engineered plants. *Science* **290**, 2088–2093.

Wraight, C.L., Zangerl, A.R., Carroll, M.J. and Berenbaum, M.R. (2000) Absence of toxicity of Bacillus thuringiensis pollen to black swallowtails under field conditions. *Proceedings of the National Academy of Sciences of America* **97**, 7700–7703.

WWF (2000) *Transgenic Cotton: Are there benefits for conservation?* URL: http: //www. panda.Org/ livingwaters

Policy Strategies for Reducing Pollution 9

9.1 INTRODUCTION

Chapter 1 identified that there are two basic options for addressing the problem of agricultural pollution. We can 'cure' the problem by acting against the chemical pollutants themselves (e.g. by water treatment), or 'prevent' the problem by addressing the underlying causes of pollution (e.g. by encouraging the adoption of alternative agricultural practices that are less polluting). Since it is generally assumed that 'prevention is better than cure', the central concern of this book has been the encouragement and pursuit of practical farm-based solutions to the avoidance of agricultural pollution. The preceding chapters have explored a number of environmental problems associated with contemporary agricultural systems. Each chapter considered the way in which agricultural management practices can be modified in order to address these pollution issues via specific measures such as:

- the application of new technologies e.g. non-chemical weed control methods (Chapter 7);
- the provision of information and advice to encourage good agricultural practice by farmers e.g. MAFF Code of Good Agricultural Practice for the Protection of Water (Chapters 2 and 3);
- statutory controls and regulations to avoid undesirable practices e.g. planning restrictions under the Town and Country Planning (General Permitted Development) Order 1995 on the location of new livestock units to reduce the risk of odour nuisance to local residents (Chapter 5).

This final chapter assesses the role of governmental policy and legislation in reducing agricultural pollution.

9.2 THE EMERGENCE OF CONTEMPORARY AGRI-ENVIRONMENTAL POLICY

Modern agricultural practice has been remarkably successful at increasing food production over the last 50 years. Rapid advances in science and technology, combined with the favourable economic climate created by post-war agricultural policies encouraged farmers to abandon traditional husbandry methods and mixed farming systems in favour of more specialist and intensive enterprises, with a greater reliance upon purchased inputs such as pesticides and fertilisers.

It is widely acknowledged that the production-orientated objectives of post-war agricultural policy are less appropriate now since, amongst other things, they have brought significant costs to bear upon the natural environment. Reform of agricultural policy and practice is continuing, with one objective being the reduction of environmental damage associated with intensive production methods. The policy agenda of most developed countries is therefore now shifting towards the integration of agricultural and environmental policy, including the encouragement of more sustainable agricultural practices (OECD, 1993, 1995, 1997; MAFF, 2000). This policy position has taken some time to develop. In the UK, for example, the environmental impacts of pesticide use were first identified in the early 1960s, but it was not until the 1980s that it became an issue of widespread public and political concern (Ward, 1996). The apparent lack of concern for so many years has been attributed to the paucity of available data on pesticide pollution. In consequence, there was no adverse public response to pesticide use and therefore no pressure upon policy makers to act.

Increasing emphasis is now being placed upon the replacement of those agricultural policies, such as price support, which encourage agricultural practices with an adverse impact upon the environment with other policies which encourage farmers to recognise that it is in their, and society's interests to maintain and enhance the environmental resources influenced by their activities (Runge, 1994). The European Commission, under Agenda 2000, has stated its aim to create a stronger agricultural industry that has as a clear focus the maintenance of the environment and the preservation of Europe's rural heritage (Brassley and Lobley, 2002). This trend effectively takes us full circle; completing the Pressure-State-Response loop (Box 9.1) suggested by OECD (1997) by returning to modify one of the most important underlying causes of agricultural pollution, namely the policies that were originally responsible for the rapid expansion, specialisation and intensification of agriculture. The emergence of environmental considerations in UK agricultural policy is also relatively recent, but developed rapidly during the 1990s. In 1979, the official UK Government view was that 'a sustained increase in agricultural net production is in the national interest and can be achieved without undue impact upon the environment' (Farming and the Nation, 1979 Government White Paper cited in Jenkins, 1990). By 1991, however, a major watershed in national policy had occurred as indicated by the strategy document, Our Farming Future, which set out Government thinking on the nature of agricultural change (MAFF, 1991; Waters, 1994). This foreshadowed the 1992 CAP reforms and outlined how the UK Government intended 'to encourage the reconciliation of agricultural and environmental objectives through an appropriate combination of advice, clear regulation and financial incentive'.

Agri-Environment Regulation 2078/92

The stated objective of the 1992 reforms, contained within the Agri-Environment regulation (2078/92) was to '.encourage farmers to use less intensive production methods, thereby reducing their impact on the environment and the creation of surpluses.' The 1992 CAP reforms represented a major shift in agricultural policy (especially in the arable, beef and sheep regimes) away from the established mechanisms of market/price support for stabilising farm incomes towards direct payments on the basis of crop area and the number of livestock. It also made funding available for the first time on a large-scale pan-European basis for an accompanying set of environmental measures under EEC Regulation No. 2078/92 which are commonly referred to as the Agri-environment Programme (EC, 1992). The Agri-environment Programme was the only specifically environmental element in the 1992 regulations, and was entirely separate from the main body of reforms. During negotiations, the UK had proposed that environmental conditions be attached to all income support payments made directly to farmers. This is known as 'cross-compliance' and was first advocated in the United States in the 1980s. However, this concept attracted little support from the European Commission and the only environmental conditions within the main body of the 1992 CAP reforms were stocking density restrictions on livestock headage payments (Baldock and Beaufoy, 1992).

Box 9.1 Pressure–State–Response (PSR) Framework

The Pressure–State–Response (PSR) framework may be employed for investigation of the linkages and feedback between economic activity and the environment (OECD, 1997). This framework considers that the relationship between contemporary economic activity and the environment functions as a simple 'loop' within which: **pressure** from human and economic activities leads to changes in the **state** or condition of resources which provoke **responses** by society to change the pressures on, and state of, the environment.

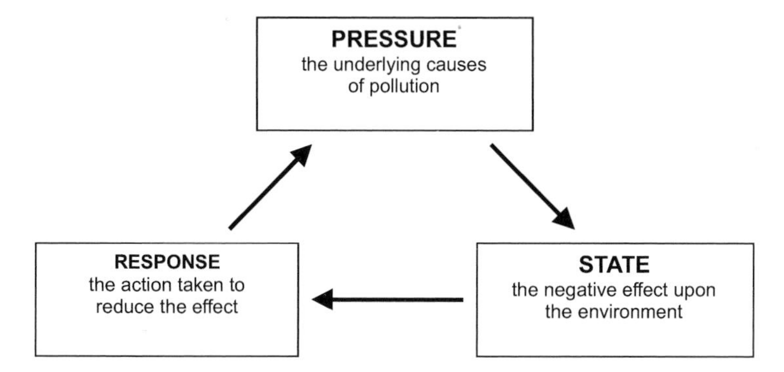

The PSR framework is very flexible with numerous variants for specific economic sectors. It is well suited to the investigation of agriculture and pollution since it offers the basis for an integrated analysis within which scientific understanding of the causes, effects and control of agricultural pollution is seen to be inextricably interwoven with the broader economic, social and political context in which agriculture operates. In particular, the PSR framework implies the fundamental importance of understanding the linkages between agricultural policy, production and pollution as the basis of developing sustainable agricultural systems.

The Agri-environment Programme made provision for EU funding of schemes proposed by Member States which encouraged farmers to adopt environmentally-sound production methods, including significant reductions in potentially polluting inputs such as fertilisers and pesticides. The total cost (payments, running costs and monitoring) of all UK agri-environment schemes, including the Environmentally Sensitive Areas (ESA) Scheme, the Countryside Stewardship Scheme and the NSA Scheme in the late 1990s was estimated to be approximately £87 million (HoC, 1997). This amount was less than 5% of the total public expenditure planned for CAP support schemes in the UK during the same period.

Rural Development Regulation 1297/1999

The Rural Development Regulation (1297/1999) has superceded previous European legislation, and aims to meet three objectives: to create a stronger agricultural and forestry sector, to improve the competitiveness of rural areas, and to maintain the environment and preserve Europe's rural heritage (Brassley and Lobley, 2002). Agri-environment measures are the only compulsory element of the Rural Development Regulation. The England Rural Development Programme (ERDP) underpins the UK Government's New Direction for Agriculture by helping farmers and foresters to respond better to consumer requirements and become more competitive, diverse, flexible and environmentally responsible (DEFRA, 2001). The scheme is regarded as highly discretionary, and promotes flexibility in the approach of individual Member States to the support of rural development activities. In Britain, the Rural Development Regulation is implemented through a series of Rural Development Plans that are specific to each country. The ERDP is based on Agri-Environment regulation 2078/92 schemes expanding or continuing, under the land-based element of the scheme, which include:

- Countryside Stewardship Scheme
- Environmentally Sensitive Areas
- Farm Woodland Premium Scheme
- Organic Farming Scheme
- Woodland Grant Scheme

It is estimated that these schemes will absorb over 70% of the spending under the EDRP between 2000 and 2007 (Brassley and Lobley, 2002). Within the ERDP it is planned that a number of the schemes introduced under the Agri-Environment regulation 2078/92 will continue the pattern of agricultural de-intensification, thus leading to a reduced risk of pollution. Evaluation of the effectiveness of these environmental schemes has begun (EC, 1998), but the impact over a longer time period remains to be tested.

9.3 POLICY INSTRUMENTS, STRATEGIES AND IMPLEMENTATION

Ultimately it is the farmers responsible for managing land at a local level who will, or will not, translate agri-environmental policy into real action with tangible results. Eckerberg and Forsberg (1996) suggested that to understand better the way in which policies influence farmers' behaviour (including the adoption of less polluting practices), it is necessary to consider some basic concepts about policy and policy making. Governmental agreements at a national and/or international level establish broad 'policy frameworks'. In order to be effective, these policy frameworks encompass three key components – a strategy (or number of strategies), instruments and implementation.

Policy strategies

Policy strategies expand upon a general policy framework by specifying first, more detailed and quantifiable policy objectives, and secondly, how these objectives will be pursued. Since it is rare for one policy instrument to achieve all policy objectives simultaneously, strategies must select the most appropriate combination of instruments to achieve optimal pollution abatement. The OECD (1991) for example, suggested that environmental policy instruments should be chosen according to five sets of criteria:

- environmental effectiveness;
- economic efficiency;
- equity;
- administrative feasibility and cost;
- acceptability.

As with most areas of policy-making, environmental policy strategies are often formulated and introduced on the basis of imperfect and incomplete information in response to perceived/unsubstantiated threats to the environment and/or public health (OECD, 1997). Rather than wait until full scientific certainty is reached about the nature and extent of this risk, prudent policy-making demands that the precautionary principle is applied and action is taken against an environmental threat on the assumption that it is 'guilty until proven innocent'.

Policy instruments

These are the means or mechanisms by which specific policy objectives are pursued. As various authors have identified (e.g. Conway and Pretty, 1991; MAFF, 1991; Eckerberg and Forsberg, 1996), the encouragement of environmentally-friendly farming methods depends upon using an appropriate 'mix' of three types of policy instrument, communicative, regulatory and economic.

Communicative instruments include the provision of information and advice as well as the opportunity for dialogue and negotiation. These instruments are used extensively in many areas of environmental policy and according to OECD (1996) their goal is to achieve the delivery of policy objectives via the simple process of 'enlightened self-interest'. For example, farmers are often advised that the use of an alternative practice is not only better for the environment, but can also save on inputs and improve business profitability. Communicative instruments are particularly important for controlling agricultural pollution because of the need for farmers to use information, management ability and ecological understanding to replace or rationalise the use of agrochemical inputs and/or other management practices (Lockeretz, 1991).

Regulatory instruments involve the 'traditional' policy mechanisms, such as statutory prohibitions and legal sanctions, that form the basis of state intervention and control in most industrialised countries. One of the principal roles of regulation in pollution control is to establish maximum ceilings or standards for acceptable levels of pollution (Conway and Pretty, 1991). This is commonly done by setting environmental quality standards for the environmental resource receiving the pollutant (e.g. drinking water standards for nitrates and pesticides). Certain agricultural practices may similarly be regulated via the market-place with the imposition of strict quality control standards and procedures upon farmers by key players, notably the multiple retailers, in the agri-food industry (Taylor and Welford, 1994).

According to Lowe and Ward (1997) the statutory regulation of agricultural pollution in the UK is not simply a technical and legislative issue, but has involved the re-orientation of traditional attitudes within the farming community to accept the sanctions and controls imposed upon their businesses. This re-orientation is necessary since until the 1980s agricultural pollution was largely a 'non-problem' with the principal agricultural imperative being production rather than environmental protection. The Control of Pollution Act 1974, for example, exempted UK farmers from prosecution for pollution if they were following 'good agricultural practice'. Once agricultural pollution was recognised as a problem, legislative action quickly followed which removed agriculture's exemption from pollution control and subjected farmers (especially livestock farmers) to much more detailed and stringent regulation and inspection, notably from the National Rivers Authority (NRA) established in 1989 as the independent regulatory body responsible for water pollution control in England and Wales. This function became the responsibility of the Environment Agency at its inception in 1996.

Section 92 of the Water Resources Act 1991 and the resulting Control of Pollution (Silage, Slurry and Agricultural Fuel Oil) Regulations aim to prevent pollution by silage effluent, dirty water, slurry and fuel oil by setting standards for keeping and handling these substances. Practical recommendations emanating from these regulations are contained within the Code of Good Agricultural Practice for the Protection of Water, which as a statutory code under Section 97 of the Water Resources Act 1997 provides advice to the agricultural community. Not adhering to the Code is not in itself an offence, but could be taken into account in any legal action (MAFF, 1998)

Economic instruments involve the use of financial incentives and disincentives to encourage or discourage the adoption or continuation of specific agricultural practices. Financial incentives are potentially very powerful instruments for modifying the behaviour of farmers – they are flexible, easily-targeted and can be linked to the implementation of both regulatory and communicative policy instruments to help achieve specific objectives (OECD, 1996). Furthermore, they are unlikely to require any re-orientation of farmers' attitudes. Examples of financial incentives include the governmental provision of subsidies, capital grants, credit or low-interest loans (Conway and Pretty, 1991), as well as the market advantage and/or premium prices obtained for certified and labelled products from environmentally-friendly farming systems (OECD, 1995).

The use of subsidies and capital grants to sproduce environmental benefits is a common form of financial incentive within agri-environmental policies. Subsidies (i.e. financial support payments) to farmers can be harnessed to environmentally-friendly practices either through cross-compliance or via voluntary management agreements

(Box 6.2) which involve regular payments for specified periods of time usually on an area basis. In contrast, capital grants normally involve one-off payments for investment in specific tasks (e.g. tree-planting) or facilities (e.g. waste handling and storage) that have environmental benefits. However, unless grant rates are 100% (i.e. none of the cost is shared by the farmers) their uptake can be limited by the reluctance of farmers to meet the additional costs over and above the grant, especially where these are perceived as producing little personal benefit (Crabtree and Chalmers, 1994).

Obviously, the success of the financial incentives outlined above at modifying the behaviour of farmers depends upon the ability and willingness of taxpayers to pay for the environmental benefits which are accrued. However, other incentives can be pursued more directly by the general public as consumers. Environmentally-friendly practices can be encouraged through the adoption of production methods according to prescribed environmental standards or codes of practice which have a strong 'market-linkage' (OECD, 1995). Accredited products with recognisable labels often have a market advantage and in some cases (e.g. organic food) may attract premium prices which significant numbers of consumers are willing to pay.

Financial disincentives are, according to Scheele (1997), the basic stimulus behind environmental protection strategy is '. . . to confront the user (or polluter) of the environment with the full economic consequences of his/her actions'. This is usually translated into the so-called 'Polluter-Pays Principle' whereby those responsible for causing the negative externalities generated by the harmful effects of economic activity upon the environment (mainly, but not exclusively, by pollution) are forced to bear the cost of this damage and/or the costs incurred in controlling the damage. In other words the external costs of economic activity are 'internalised' to become part of the normal business costs, thereby theoretically encouraging the adoption of less harmful practices/technologies.

The 'Polluter-Pays Principle' is very well-established in environmental policy (OECD, 1975) and is most commonly applied in agriculture via the government imposition of taxes on fertilisers and pesticides (Conway and Pretty, 1991). However, studies suggest that if significant reductions in the use of these inputs are to be made then very high taxes (e.g. well in excess of 200% of the market price) are required (OECD, 1989). No policy-makers have yet attempted to introduce such drastic 'supply control' taxes, preferring instead to impose relatively small revenue-raising 'environmental' taxes which generate funds for investment in research or extension services. Although, as Clunies-Ross (1993) points out, this approach does risk enshrining the right to carry on polluting by encouraging farmers to pay the tax as an acceptable additional cost rather than to alter their practices. A further criticism of taxing agrochemical inputs as a means of pollution control is that the incidence of pollution on individual farms is influenced by many other environmental factors and husbandry practices. Equally there is no incentive for farmers to adopt 'good agricultural practice' if they will continue to be penalised on the same basis as other farmers who ignore good practice.

A better approach (assuming an appropriate mechanism can be found) may be to impose a tax or levy payment upon pollution itself. The Dutch government implemented legislation in 1987–1988 that included the introduction of a levy system that charges farmers for producing surplus manure on their farms (Tamminga and Wijnands, 1991). Although innovative, the success of a system such as this depends upon first, the participating farmers being sufficiently competent in the collection, management and processing of relevant data and secondly, farmers having sufficient income/motivation to afford the extra time and expense involved in monitoring manure production on their farms; and thirdly the government having the means to monitor farmers' activities and to detect violations (OECD, 1995). The range of financial disincentives available to control pollution are discussed in more detail by various authors in Romstad *et al.* (1997).

Implementation and dissemination

This is the organisational arrangements within which policy strategies are implemented. The stakeholders within this structure may include farmers and their representatives, governmental agencies, sector authorities, private interest groups and even the general public, while their success at implementing policy will depend upon:

- the way in which they organise themselves to solve problems of policy implementation;
- their degree of power and authority;
- the level of resources the stakeholders are allocated.

The implementation structure will obviously vary depending upon the policy strategies and instruments adopted. For example, according to Eckerberg and Forsberg (1996) regulatory instruments tend to be associated with centralised decision-making and 'top-down' policy implementation. Communicative instruments on the other hand aim to encourage decentralised decision-making and 'bottom-up' policy implementation by developing common knowledge and understanding between the policy makers and individual farmers, and leaving the final decisions on specific management practices and actions to the individual farmer.

As a general principle, environmental policy strategies and their implementation structures should be developed with a view towards minimising as much as possible the public costs of administration, monitoring and enforcement (OECD, 1996). One low-cost approach to implementing environmental policy which is increasingly favoured in some countries is the government funding of voluntary and community assistance programmes to build the 'capacity' of local people to address local environmental problems with locally-developed solutions.

Australia, Canada and New Zealand in particular have placed great emphasis upon the development of self-reliant, community-based approaches to agricultural resource management (OECD, 1995). For example, about 25% of Australian farmers in some states belong to Landcare groups, a range of publicly and privately funded programmes developed in response to local agri-environmental problems. According to Alexander (1993), 'Landcare is a grass-roots movement of groups of farmers working together with government and the wider community to solve rural land use problems . . . it is the epitome of the education and persuasion rather than legislation and coercion approach to environmental protection'.

Although other countries have yet to move as far as Australia in devolving responsibilities for environmental protection to local groups, many are finding that agri-environmental policy development and implementation is much more effective when the farmers themselves are involved in the process. Since the introduction of the EC Agri-environment Programme and the encouragement of locally-distinct 'zonal programmes', many EU Member States have recognised that the long-term success of agri-environmental schemes does not just depend upon technical knowledge and expertise, but also issues of greater accountability and sensitivity to farmers and their local communities (Baldock and Mitchell, 1995).

Research conducted for MAFF in England and Wales in relation to the Nitrate programme has shown that farmers view one-to-one interaction, written text and farm demonstrations as the most effective means of communicating technical information related to nitrate losses from farmland. Peer group example and pressure were also reported as being effective (Dampney *et al.*, 2000). Some elements of the dissemination process in relation to agricultural pollution has been regarded as particularly successful. For example, Robinson (1999) argued that the reduction of point source pollution incidents in England and Wales in the 1990s was due to the implementation of the Control of Pollution (Silage, Slurry and Agricultural Fuel Oil) Regulations as part of a package which included publicity (underpinned by a clear prosecution policy by the former NRA), development of technical solutions (particularly relating to silage effluent and slurry storage), introduction of regulations including technical specifications and provision of grants to assist farmers to improve manure and silage storage infrastructure on farm. It is suggested that this approach might provide a model for successfully promoting change in other areas within the agricultural sector.

In contrast, the Code of Good Agricultural Practice for the Protection of Water when first introduced in 1991 received less attention from the farming industry than was intended. Certainly it has been pointed out by Dampney *et al.* (2000) that advice targeted at a range of issues, many of them diffuse in origin, such as nitrates, is less well received by farmers. Recent revisions of all the Codes of Good Agricultural Practice have focussed on a clearer presentation of the issues. That, and a progressive realisation within the industry that environmental quality issues are 'here to stay' has led to a more widespread adoption of advice contained within the Codes. However, there is still a long way to go before the majority of the farming community not only accept the Codes, but put the guidance into practice on-farm.

9.4 LEARNING FROM EXPERIENCE

The remainder of this chapter looks at two examples of the policy strategies already in place for reducing agricultural pollution. It does not aim to provide detailed prescriptions, but rather to illustrate key developments in policy-making (including both successes and intractable problems) as a source of ideas and insight into the control of agricultural pollution. The areas examined are those specific policies for the control of pollution associated with nitrates in water and pesticide contamination of soil and water.

Policies for reducing nitrate pollution

Nitrate pollution of ground, surface and marine waters is a major environmental issue in many countries (with potential implications for human health) which has attracted considerable public and political attention. In contrast to pesticides, however, the control of nitrate pollution is not simply a question of rationalising and reducing an agrochemical input. According to Watson *et al.* (1996), '. . . the nitrate problem is characterised by complexity, uncertainty and conflict'. As Chapter 2 explained:

- Nitrate leaching is an example of diffuse pollution which arises from a variety of sources and is influenced by a number of factors, including interactions between the physical, chemical and biological components of the agro-ecosystem and the wider agricultural economy. For example, while the steady increase in the use of nitrogen (N) fertilisers is undoubtedly significant, the precise relationship between the rate of N applied and the incidence of nitrate leaching is not always clear and direct.

- The public health hazards and ecological disruption associated with nitrate pollution are still subject to debate. For example, the link between infantile methaemoglobinaemia (blue baby syndrome) and very high levels of nitrate intake in bottle-fed babies is well documented, but very rare, while there is little consensus amongst experts on the potential link between nitrates and gastric cancer. There is also considerable uncertainty about the relative importance of nitrate and phosphate in causing eutrophication in surface and marine waters.

Due to the complexity and uncertainty of the nitrate issue, conflicts have inevitably arisen among agricultural, environmental, governmental and industrial interests regarding the severity of nitrate pollution and the nature of, and responsibility for, the actions that need to be taken. These conflicts have arisen at national, European and international level and many authors have suggested that nitrate pollution is a political issue as much as an environmental problem. As Elworthy (1994) noted on the situation in the UK, 'the dearth of scientific knowledge about nitrate leaching has, arguably, left a vacuum in which a furious political battle has raged between, on the one side, the agrochemical industry, farmers, the water industry and the government and on the other environmental groups and a number of individual scientists'.

Nonetheless, it is widely acknowledged that a major factor contributing to the increased pollution of the aquatic environment by nitrates has been the specialisation and intensification of agricultural enterprises; notably the decline in traditional mixed farming systems (i.e. crops and livestock on the same farm) and the increase in specialist arable and livestock farms. These specialist production enterprises are more susceptible to nitrate leaching due to:

- the accumulation of N surpluses within the farming system. These surpluses increase with the intensity of agricultural production, although their occurrence and extent differs considerably between farming types and different countries (Brouwer and Hellegers, 1997);

- changes in land management practice which increase the risk of surplus nitrate 'leaking' from the agro-ecosystem into the aquatic environment (Addiscott *et al.*, 1991).

In contrast to the control of point source pollution (e.g. laws prohibiting the direct discharge of livestock wastes into surface waters), the favoured policy on nitrate pollution for many years was to place maximum allowable limits on drinking water and thereby oblige the water industry to remove/dilute the nitrate before delivery to the consumer (Conway and Pretty, 1991). Most developed countries have now introduced specific policies to control nitrate pollution by encouraging farmers to reduce the N inputs on their farms and/or modify other management practices which affect nitrate leaching. However, it should be noted that while most national nitrate policies involve some form of direct regulation of nitrogen inputs (fertiliser and manure), in some cases these were originally intended to reduce surplus production rather than protect water resources. For example, there has been a levy on fertiliser use in Austria since 1986 which was introduced originally to curb surplus production. However, since a 1991 amendment to the Austrian Federal Act on Water this levy is now part of a policy strategy which aims to control nitrate leaching through 'orderly agricultural practice' (OECD, 1993). This new legislation, which works on a province-by-province basis, also imposes limits upon stocking rates.

Nitrate pollution control policies in the UK and Europe

The quality of drinking water is one of the most important environmental issues in Europe and nitrate pollution control policies have been in place for many years. For example, EC legislation setting controls on the level of nitrates in drinking water was first introduced in 1975, but at this time only extended to cover abstraction from surface water. More comprehensive legislation involving a total of 62 water quality standards (including nitrates) was introduced in the 1980 EC Drinking Water Directive (EC, 1980). This was intended to protect human health at the point of supply to the consumer and full compliance with the Directive was required amongst all EU Member States by mid-1985.

National nitrate pollution control policies and schemes have varied across the EU according to the dominant issues in individual countries. For example, while Germany placed particular importance upon localised water protection zones to reduce the nitrate contamination of boreholes and wells (Conway and Pretty, 1991), Denmark and the Netherlands focused upon national controls over the production and use of manures (PARCOM, 1993). According to Frederiksen (1995), most national nitrate pollution control policies introduced in EU Member States prior to 1993 involved voluntary restrictions and codes of good agricultural practice; regulations concerning the application and management of manure and fertiliser N on the farms, and economic incentives directed towards subsidising farmers introducing environmentally-friendly practices (levies on excess manure production were also applied in Belgium and the Netherlands). The broad categories of regulatory instrument used in national policies to control N application and management were:

- rules relating to the amount of N which can legally be applied per hectare per year, either as a total amount of manure (i.e. leaving the possibility open for application of additional fertiliser N) or as a total of both manure and fertiliser;
- regulations concerning when, how and where manure can be applied;
- requirements for minimum storage capacity for manure and slurry.

The development of more rigorous nitrate control policies within EU Member States has largely been in response to the legislative framework imposed by the European Commission since 1980, including:

- Drinking Water Directive (EC, 1980);
- Nitrates Directive (EC, 1991);
- Agri-environment Programme (EC, 1992);
- Rural Development Regulation (EC, 1999).

Recent policy in the UK has been dominated by UK government interpretation and implementation of the EC Nitrate Directive 91/676, which required Member States to introduce a Code of Good Agricultural Practice to control nitrate loss which all farmers should apply on a voluntary basis. The objective of this Directive was to reduce water pollution by nitrates from agricultural sources, and to prevent further such pollution. In addition, NVZ were designated under the Directive. Within these zones, 68 of which had been designated by 1999, an Action Plan for the reduction of pollution by nitrates is implemented. Only catchments where nitrate concentration in sources of drinking water already exceeds the EC standard, or is likely to by 2010, have been designated, although the Directive allows for the designation of NVZs in catchments where eutrophication is a major risk. Details of the practical steps included in the Action Plan for the reduction of pollution by nitrates have been discussed in Chapter 2. In outline, the regulations limit applications of fertilisers to crop requirements, impose limits on the timing of fertiliser application, and impose limits on the rates and timing of application of organic manures. In addition there is a requirement to have sufficient storage for farm wastes to cover closed periods for spreading and to keep relevant records, including numbers of livestock grazing specific areas of land. Evaluation of the effectiveness of these measures will only be possible after the NVZ scheme has been running for a number of years.

In addition to the NVZ scheme, the NSA scheme has been running since 1990 to help protect key groundwater sources. This scheme, implemented under the Water Act 1989, is voluntary, and provides for compensated measures to encourage farmers to convert arable land to extensive grassland, as well as other practices going well beyond the Code of Good Agricultural Practice for the Protection of Water. The scheme, which was closed to new applications in 1998 is due to run until 2005. A review of the initial impacts of the scheme by Lord *et al.* (1999) was discussed in detail in Chapter 2.

In Denmark, the eutrophication and pollution of groundwater due to leaching of nitrates has been a major concern since the 1980s (Dubgaard, 1991; Daugbjerg, 1998), as nitrate concentrations in Danish watercourses have been observed to be at least seven times higher in agricultural areas than in uncultivated, semi-natural zones. As a consequence, one of the main targets of the Agri-Environment measures introduced under Regulation 2078 has been the reduction in the use of nitrates. To obtain payments through this measure, farmers must reduce N inputs to 60% of that recommended by the Danish Ministry of Agriculture (Andersen *et al.,* 2000). The whole farm must be covered by the agreement. While only 6400 ha was entered directly under this scheme in 1996, a number of other schemes, such as the maintenance of extensive grassland showed much wider take-up (48,700 ha), and will also offer benefits in terms of reduced nitrate losses from farmland. Many other EU Member States have broadly similar schemes directed at changing farming practices, for example conversion to organic farming and extensification of arable land. Most of these schemes will contribute to the goal of reduce nitrate losses from agricultural land. In many cases, more time needs to elapse before the full impact of land management changes can be assessed (Buller *et al.*, 2000).

Policies for reducing pesticide pollution

Early government policies towards pesticides were mainly concerned with protecting farmers and farmworkers from the misuse of hazardous chemicals. It was not until the late 1960s and early 1970s that a range of regulatory instruments were implemented in most countries (OECD, 1993), including:

- registration schemes for the marketing and use of individual products;
- controls and recommendations over the use of pesticides;
- safety and environmental standards for the manufacture, distribution and application of pesticides;
- standards for permitted levels of residues in food;
- controls over the disposal of waste pesticides.

These instruments were not initially intended to reduce the application of pesticides and total pesticide use continued to rise as a result. It was not until the 1980s that public concern over the possible effects of pesticides upon human health and the environment (plus improved scientific methods for detecting pesticide concentrations in the environment) led to the introduction of more specific government policies on:

- withdrawing approval for the use of those pesticides that pose the greatest threats to public health and the environment;
- reducing the use of, and pollution from, those pesticides that remained approved for use (OECD, 1995).

Although some countries, such as Greece and Portugal, are still principally concerned with improving the application and storage of pesticides on farms (OECD, 1993), others are pursuing greater integration of environmental and public health concerns into pesticide policy. An important step in many countries has been the tightening of pesticide registration standards combined with legislation requiring the periodic review and/or re-registration of pesticides. If a pesticide is deemed not to satisfy contemporary health and environmental standards, it must be withdrawn from sale upon termination of the prevailing certificates of approval.

Sweden completed a 5-year re-registration programme in 1995 during which 21 active ingredients were banned (e.g. aldicarb, carbaryl, dinocap, thiram and trifluralin), 18 had additional restrictions imposed upon their use and 18 were voluntarily withdrawn (e.g. bromoxynil and 2,4-D). In all, 250 formulated products were withdrawn from use out of a total of 600 previously registered (Matteson, 1995).

A further example of regulation of pesticide use is related to their presence in water. The 1976 EC Dangerous Substances in Water Directive and 1980 EC Groundwater Directive together set a framework for the reduction or elimination of pollution by particularly dangerous substances in surface (inland, coastal and territorial marine) and ground waters (Beaumont, 1992). The dangerous substances listed in the Directives include some pesticides, principally the persistent organochlorines such as aldrin, dieldrin, DDT, lindane and chlordane. In addition to this Europe-wide legislation, most EU Member States have their own national legislation for addressing point source pollution.

Maximum Admissible Concentrations (MACs) for pesticides in drinking water were set in the 1980 EC Directive on Drinking Water Quality (Beaumont, 1992). This came into force in 1985 and specifies that the MAC for any one pesticide is 0.1 $\mu g\ l^{-1}$ (0.1 parts per billion) and 0.5 $\mu g\ l^{-1}$ (0.5 parts per billion) for total pesticides, irrespective of their individual toxicity. Several EU Member States have been forced to take action to reduce pesticide levels in drinking water, both through water treatment and programmes aimed at reducing pesticide use by farmers.

In the UK during 1999 it was reported that 15 types of pesticides had reached concentrations above the 0.1 $\mu g\ l^{-1}$ drinking water limit. However, long-term trends indicate improvement; 3% of all drinking water samples contravened the limit in 1992 which has now decreased to 0.01% (DWI, 2000).

The focus of current water quality legislation and monitoring is upon the protection of drinking water supplies and human health and it has been demonstrated that quality standards can be insufficient to protect aquatic wildlife (HMSO, 1995). The implementation of Environmental Quality Standards (EQSs) is an attempt to set limits for substances that can cause environmental harm. For aquatic environments, a given EQS should not be exceeded in order to protect that water bodies (Environment Agency, 2000). Values are set for each substance based on toxicological information; there are currently 66 EQSs for pesticides.

Pesticide reduction programmes in Europe

Sweden, Denmark, and the Netherlands adopted pesticide reduction programmes in the late 1980s in response to two key issues. First, increasing public concerns about agrochemical pollution supported by environmental monitoring showed that the concentration of pesticides in the upper layer of groundwater beneath 65% of agricultural land in the EU exceeded drinking water standards (Reus *et al.*, 1994). A large number of pesticides also exceed or are expected to exceed existing standards for surface water. Secondly, there was a perceived need to keep the agricultural sectors as competitive as possible. In particular because the viability of domestic agriculture and the export of agricultural products (especially vital to agricultural viability in Denmark and the Netherlands) were felt to be increasingly threatened by:

- the overuse of expensive agrochemicals;
- pesticide resistance problems;
- the implementation of Maximum Acceptable Concentrations for pesticide residue levels, as well as other pesticide-related export restrictions.

The general aim of the pesticide reduction programmes in all three countries has been to maintain agricultural productivity while meeting national commitments to reduce the total quantity of all pesticides used in agriculture by at least 50% before the year 2000 (Table 9.1). Sweden initiated its Pesticide Risk Reduction Program in 1986 with the aim of reducing the quantity of pesticide active ingredients used by 1990 to 50% of average use during the base period 1981–1985. Similar targets were set in Denmark under the Action Plan to Reduce Pesticide Application (50% reduction by 1990) and in the Netherlands under the Multi-Year Crop Protection Plan (50% reduction by 2000).

Table 9.1 National Pesticide Reduction Targets in Sweden, Denmark and the Netherlands (Hurst *et al.,* 1992; OECD, 1995, 1997).

	National Programme	Target(s)
Sweden	Pesticide Risk Reduction Program 1986–1997	50% reduction in total agricultural use of pesticide active ingredient by weight from 1986–1990. A further 50% reduction was mandated for 1991–1997 (i.e. a total reduction of 75%)
Denmark	Action Plan to Reduce Pesticide Application 1986–1997	50% reduction in total agricultural use of pesticide active ingredient by weight. 50% reduction in pesticide application frequency (defined as the average number of label rate dosages applied annually to cultivated land)
The Netherlands	Multi-Year Crop Protection Plan 1990–2000	50% reduction in total crop agricultural use of pesticide active ingredient by weight. Reduced dependence on pesticides (not quantified). Reductions of emissions of pesticide active ingredient by weight: 50% to air; 75% to ground water/'non-target' soil; 90% to surface water

These latter programmes also included targets to reduce the frequency of pesticide application. In Denmark, regulations restrict the number of pesticide applications per growing season, while in the Netherlands growers have been forbidden since 1993 to apply soil fumigants more than once every four years (OECD, 1995). Soil fumigants were extensively used in potato, onion and flower bulb production and accounted for over 50% of all pesticides used in the Netherlands in 1990. It is intended that these additional restrictions on the frequency of application will force farmers to increase their reliance on alternative pest management methods, including the use of more balanced crop rotations. The Dutch government also adopted a third target that called for an elimination of most pesticide emissions to air, 'non-target' soil and water by the year 2000. This was in recognition of the very high pesticide use characteristic of agriculture in the Netherlands, and the shallow water table and numerous lakes and watercourses which make the Dutch national water supply particularly vulnerable to pollution.

The policy instruments used to pursue the reduction targets within these national programmes have included a combination of:

- further tightening of pesticide registration standards to eliminate or restrict the use of particularly hazardous chemicals;
- mandatory buffer zones (i.e. areas where pesticides may not be applied) to reduce the environmental impact and total volumes of pesticides used;
- mandatory training of pesticide users and the testing/certification of spraying equipment to ensure that safety standards are met, necessary records are kept and only the most efficient methods of application are used;
- greater collaboration in policy making and implementation between the government agencies that administer the programmes and farmers' organisations and trade associations. These links have also been used to increase the flow of information about the fate of pesticides in the environment and to identify and promote more sustainable agricultural technologies and practices;
- the availability of grants to help farmers cover the expense of adopting environmentally-friendly practices, such as organic farming or the establishment pesticide-free buffer zones adjacent to ecologically sensitive areas (Hurst *et al.,* 1992; OECD, 1995).

The Swedish and Danish programmes are paid for by pesticide taxes, while the cost of the programme in the Netherlands is shared by the government and agricultural industry under a covenant signed in 1993. Matteson (1995) noted that there have been successes with the pesticide reduction programmes in all three countries. In 1990, following the successful reduction of pesticide use by 47% (without any decline in average crop yields) the Swedish parliament introduced more restrictions to reduce pesticide use by a further 50% to 25% of the average use in the period 1981–1985. Similar successes in progress towards the 50% reduction target have been reported in Denmark and the Netherlands, but achieving reductions in the frequency of application (notably in Denmark) has been more difficult. This seeming paradox is apparently explained because reductions in pesticide use have mainly been achieved by improving the efficiency of application (i.e. using less active ingredient per application), rather than substituting pesticide application with non-chemical pest control methods (Matteson, 1995). In response to this, the Danish government has implemented greater restrictions on pesticide applications, mandatory record keeping and more attractive subsidies for low input farming systems. National pesticide taxes have also been increased to provide more funds for environmental schemes and to increase research on organic farming.

These examples from northern Europe are evidence of the potential to apply pesticide reduction programmes without compromising the well-being of a country's agricultural sector. According to Mattesson (1995), some of the key lessons to be learnt are:

- The importance of setting measurable targets with strict deadlines in order firstly to stimulate action amongst farmers, researchers and advisors, and secondly to preserve momentum in a goal-orientated pesticide reduction process.

- The need to be cautious and choose realistic initial programme targets that have a high chance of being achieved so that the programme is perceived as a success and thereby worthy of further public and political support. For instance, both Sweden and Denmark have greatly reduced the quantity of pesticides used in agriculture, but have yet to reduce the frequency of application. However, whilst Sweden's 'Pesticide Risk Reduction Program' is a great success according to the targets it set for itself, Denmark's 'Action Plan to Reduce Pesticide Application' appears not to be.

- The need to monitor the progress being made towards programme targets in order that policy makers can identify weaknesses and gaps in the reduction programme and take appropriate measures to improve its performance (e.g. allocate resources to new policy instruments).

- The importance of fully integrating the principles of pesticide reduction into all agricultural policy. In some instances in northern Europe, for example, seemingly unrelated policies and regulations which are valuable in another context have inadvertently promoted, continued, or even increased, pesticide use.

Policies that have conflicted with pesticide reduction include first, the privatisation of agricultural extension in Denmark and the Netherlands, making access to advice and services which were previously free more difficult and expensive. This discouraged farmers from seeking technical assistance and had a negative impact on pesticide use reduction because alternative pest management practices are information and knowledge 'intensive'. Secondly, the discontinuation of funding in Denmark for a management economics programme developing techniques for farmers' evaluation of new pest management technology. Thirdly, environmental protection measures in Denmark promoting an expansion of winter cropping to reduce nitrate leaching caused an increase in pesticide use because winter wheat crops require more pesticides than spring barley (Hurst *et al.*, 1992).

9.5 FUTURE DIRECTIONS

Adoption of sustainable farming practices

During the last 10 years it has become apparent that there is a powerful case to be made for linking production practices in agriculture with protection of the rural environment. Scientific evidence, much of it reviewed in this book, now clearly demonstrates the link between inputs, management practices and outputs from agricultural systems and their consequent environmental impact. The widespread debate stimulated by governmental and non-governmental organisations regarding the appropriate use of resources and land management practices is also driving the development of environmentally sensitive farming practice. For example, LEAF (Linking Environment and Farming) is one of a number of non-governmental organisations throughout Europe that actively encourages the adoption of practical, integrated crop and farm management (ICM and IFM). DEFRA now encourages discussion and adoption of target indicators for a more sustainable agriculture (MAFF, 2000) which include conventional measures of environmental quality such as nutrient and pesticide concentrations in rivers, as well as more general indicators of system health, such as soil organic matter content. This process is being enhanced by the major retailers who, through their quality protocols, encourage farmers to adopt more environmentally friendly, less polluting farming practices.

To this end, the concept of increased sustainability within agriculture has emerged as an important guiding principle that is shaping the outlook of governmental and non-governmental organisations around the world. There is now a widespread consensus that food production systems are needed which are not only capable of feeding a growing population and generating sufficient returns to maintain agricultural communities, but can also sustain these objectives by:

- avoiding heavy reliance upon any external inputs, notably those utilising non-renewable resources, such as fossil fuels;
- maintaining and improving the means of production;
- producing minimal negative impact upon the natural environment.

The widespread adoption of agri-environment schemes throughout Europe following the introduction of Regulation 2078/92 has made a significant step towards sustainability (EC, 1998). By 1998 agri-environment contracts that deliver environmental services applied to over 20% of European farmland. There is much evidence, summarised by the EC (1998) that shows the substantial benefits that have accrued from these agri-environmental programmes. Many of these directly or indirectly lead to reduced pollution potential, for example:

- many of the broad programmes to limit inputs use significantly less N-fertiliser and better application techniques, resulting in a lower N surplus;
- strong evidence of positive activities for nature protection is available. This results from the symbiotic link between European biodiversity and farming on which it depends;
- application of programmes has had some effects on changing attitudes, both of farmers and the general public. Farming is increasingly seen as an activity consistent with environmental care.

Further monitoring and assessment is underway across Europe to evaluate the longer term environmental benefits of the Agri-Environment and the Rural Development Regulation.

Legislative changes

The implementation of the Rural Development Regulation will continue the shift from production subsidies to support for farming practices conducted in sympathy with the wider environment. It is clear that scientific research has identified a number of solutions to pollution problems that are caused by 'inappropriate' land management practices. Furthermore, the farming industry now largely accepts that it has a responsibility to manage land in a more environmentally benign manner. However, the process of technology transfer still remains frustratingly slow in some cases. In addition, as was noted by EC (1998) there has been less progress in regions where intensive farming produces high value crops. In these cases, yield reductions are often too substantial for price support mechanisms to be able to fully compensate for lost profit associated with reduced inputs. In such cases a stronger regulatory framework is more likely to achieve effective pollutant reduction targets.

Several key articles of national government legislation have had a direct impact on pollution issues discussed in this book. For example, the Water Act 1989 included the legislation that led to the establishment of the NSA scheme. Similarly, statutes in the Environmental Protection Act 1990 led to the ban on straw burning in 1993. The Environment Agency, the government agency that takes the lead in monitoring pollution in the rural environment, was set up following the enactment of the Environment Act 1995. Further legislation will follow recent European directives. The introduction of the IPPC Directive (96/91/EC) represents a positive step to control all potential emissions from industry. The main purpose of the IPPC Directive is to achieve integrated prevention and control of pollution from listed activities. This is to be done by preventing, or where that is not practicable, reducing emissions to the air, water and land by potentially polluting industrial and other installations 'so as to achieve a high level of protection of the environment taken as a whole' (NSCA, 2001). Within the directive, emphasis is placed in identifying 'best available technology' and encouraging industry to adopt production technologies that provide an acceptable balance between minimising environmental emissions and other criteria including cost (Robinson, 1999). In the case of the agriculture industry, the directive affects intensive pig units with more than 750 sows or 2000 weaners, and poultry units with more than 40,000 birds. The Pollution Prevention Control Act 1999 enables regulations to be made implementing the IPPC Directive. The Pollution Prevention and Control Regulations 2000 (Statutory Instrument No. 1973) fulfils the requirement of the IPPC Directive, and is significant in that for the first time controls have been introduced over emissions to the atmosphere from agricultural activities.

Further EU legislation is included within the Water Framework Directive (2000/60/EC), agreed in October 2000, and will be progressively implemented in Member States over the next 5 to 10 years. The Directive sets the overall goal for surface waters as good ecological status, i.e. it is an ecosystem-based approach, not one based on individual chemical parameters. It also requires integrated management of waters and full inclusion of stakeholders in the process. The Directive also requires that abstraction of surface and ground waters be managed to a sustainable level. The directive does not include any specific pollution control powers, but will be an important vehicle for the control of eutrophication. The directive will provide a framework for a coordinated approach to sustainable water management on a catchment basis. Thus it will focus attention on the issue of how to control diffuse pollution from agricultural sources, and may lead to consideration of more specific legislation at national level to address diffuse pollution (Robinson, 1999).

While legislation will continue to push forward developments and speed up the adoption of less polluting farming practices, effective steps can only be made with the full cooperation of the farming community. As Webb *et al.* (2001) commented, in a review of the impact of farming practices on soil fertility, 'reduction in pollution is, for the most part, more likely to be achieved by changing practices to conserve potential pollutants in the soil, rather than across-the-board input reductions which may have little effect on losses but cause significant reductions in crop yields'. Effective control of pollution from agricultural sources will depend on demonstrating both the environmental and economic benefits to land managers. Only then will there be a widespread adoption of more resource efficient farming practices.

REFERENCES

Addiscott, T.M., Whitmore, A.P. and Powlson, D.S. (1991) *Farming, Fertilizers and the Nitrate Problem.* CAB International, Wallingford.

Alexander, H. (1993) *Lessons in Landcare: Australia's Model for a Better Farming Future.* The SAFE Alliance, London.

Andersen, E., Henningsen, A. and Primdahl, J. (2000) Denmark: implementation of new agri-environmental policy based on Regulation 2078. In: *Agri-environmental Policy in the European Union* (Eds H. Buller, G.A. Wilson and A. Holl). Ashgate, Aldershot, pp. 31–50.

Baldock, D. and Beaufoy, G. (1992) *Plough On! An Environmental Appraisal of the Reformed CAP.* World Wide Fund for Nature (UK), Goldalming.

Baldock, D. and Mitchell, K. (1995) *Local Influence: Increasing Local Involvement in the Development of Green Farming Schemes.* Council for the Protection of Rural England, London.

Beaumont, P. (1992) Pesticides, Europe and The Environment: A Review. *Pesticide News* **16**, 12–15.

Brassley, P. and Lobley, M. (2002) The Common Agricultural Policy of the European Union. In: *The Agricultural Notebook* (Ed. R.J. Soffe). 20th Edition. (In press).

Brouwer, F. and Hellegers, P. (1997) Nitrogen Flows at Farm Level across European Union Agriculture. In: *Controlling Mineral Emissions in European Agriculture* (Eds E. Romstad, J. Simonsen and A. Vatn). CAB International, Wallingford, pp. 11–26.

Buller, H., Wilson, G.A. and Holl, A. (2000) *Agri-Environmental Policy in the European Union.* Ashgate, Aldershot.

Clunies-Ross, T. (1993) Taxing nitrogen fertilisers. *The Ecologist* **23**, 13–17.

Conway, G. and Pretty, J. (1991) *Unwelcome Harvest: Agriculture and Pollution.* Earthscan Publications, London.

Crabtree, J.R. and Chalmers, N.A. (1994) Economic evaluation of policy instruments for conservation: standard payments and capital grants. *Land Use Policy* **11**, 94–106.

Dampney, P.M.R., Lord, E.I. and Chambers, B.J. (2000) Development of improved advice for farmers and advisers. *Soil Use and Management* **16**, 162–166.

Daugbjerg, C. (1998) Linking policy networks and environmental policies: nitrate policy making in Denmark and Sweden 1970–1995. *Public Administration* **76**, 275–294.

DEFRA (2001) England Rural Development Programme. http://www.defra.gov.uk/erdp/schemes/schemesindex.htm

Dubgaard, A. (1991) *The Danish Nitrate Policy in the 1980s.* Statens Jordbrugsokonomiske Institut Report No. 9, Copenhagen.

DWI (2000) *Overview of Water Quality in England and Wales, Drinking Water 1999.* Drinking Water Inspectorate, Department of the Environment, Transport and the Regions, London.

EC (1980) Council Directive (EEC) No. 778/80 relating to the quality of water intended for human consumption, *Official Journal of the European Communities* No. **L229** (30/8/80), 1–8.

EC (1991) Council Directive (EEC) No. 676/91 concerning the protection of waters against pollution caused by nitrates from agricultural sources, *Official Journal of the European Communities* No. **L375** (31/12/91), 1–7.

EC (1992) Council Regulation (EEC) No. 2078/92 on agricultural production methods compatible with the requirements of the protection of the environment and the maintenance of the countryside. *Official Journal of the European Communities* No. **L215** (30/7/92), 85–90.

EC (1998) State of application of Regulation (EEC) No. 2078/92: Evaluation of Agri-Environment Programmes. DGVI Commission Working document VI/7655/98. European Commission, Brussels.

EC (1999) Council Regulation (EEC) No. 1257/1999 on support for rural development from the European Agricultural Guidance and Guarantee Fund (EAGGF) and amending and repealing certain Regulations, *Official Journal of the European Communities* No. **L160** (26/06/99), 80–101.

Eckerberg, K. and Forsberg, B. (1996) Policy Strategies to Reduce Nutrient Leaching from Agriculture and Forestry and their Local Implementation: A Case Study of the Laholm Bay, Sweden. *Journal of Environmental Planning and Management* **39**, 223–242.

Elworthy, S. (1994) *Farming for Drinking Water–Nitrate Pollution of Water: An Assessment of a Regulatory Regime*. Avebury Studies in Green Research, Aldershot.

Environment Agency (2000) *Pesticides 1998*. A summary of monitoring of the aquatic environment in England and Wales. Environment Agency, Bristol.

Frederiksen, B.S. (1995) National Responses to the EC Nitrate Policy. *Journal of Environmental Planning and Management* **38**, 253–263.

HMSO (1995) *Pesticides in Surface Waters*. HMSO, London.

HoC (1997) *Environmentally Sensitive Areas and other Schemes under the Agri-environment Regulation*. Second Report from the House of Commons Select Committee on Agriculture Inquiry: Volume I–Report. House of Commons Paper **45**. HMSO, London.

Hurst, P., Beaumont, P. Jorgensen, C.E. and Winther, S. (1992) *Pesticide Reduction Programs in Denmark, the Netherlands, and Sweden*. World Wide Fund for Nature International, Gland.

Jenkins, T.N. (1990) *Future Harvests – The Economics of Farming and the Environment: Proposals for Action*. A report to CPRE and WWF(UK), Council for the Protection of Rural England, London.

Lockeretz, W. (1991) Information requirements of reduced-chemical production methods, *American Journal of Alternative Agriculture* **6**, 97–103.

Lord, E.I., Johnson, P.A. and Archer, J.R. (1999) Nitrate sensitive areas: a study of large scale control of nitrate loss in England. *Soil Use and Management* **15**, 201–207.

Lowe, P. and Ward, N. (1997) The Moral Authority of Regulation: The Case of Agricultural Pollution. In: *Controlling Mineral Emissions in European Agriculture* (Eds E. Romstad, J. Simonsen, and A. Vatn) CAB International, Wallingford, pp. 59–71.

MAFF (1991) *Our Farming Future*. MAFF Publication No. 0703. Ministry of Agriculture, Fisheries and Food Publications, London.

MAFF (1998) *Code of Good Agricultural Practice for the Protection of Water*. Ministry of Agriculture, Fisheries and Food, London.

MAFF (2000) *Towards sustainable agriculture*. Ministry of Agriculture, Fisheries and Food, London.

Matteson, P.C. (1995) The '50% Pesticide Cuts' in Europe: A Glimpse of Our Future? *American Entomologist* **41**, 210–220.

NSCA (2001) *Pollution Handbook 2001*. National Society for Clean Air and Environmental Protection, Brighton.

OECD (1975) *The Polluter-Pays Principle: Definition, Analysis and Implementation*. Organisation for Economic Co-operation and Development, Paris.

OECD (1989) *Agricultural and Environmental Policies – Opportunities for Integration*. Organisation for Economic Co-operation and Development, Paris.

OECD (1991) *Environmental Policy: How to Apply Economic Instruments*. Organisation for Economic Co-operation and Development, Paris.

OECD (1993) *Agricultural and Environmental Policy Development: Recent Progress and New Directions*. Organisation for Economic Co-operation and Development, Paris.

OECD (1995) *Sustainable Agriculture: Concepts, Issues and Policies in OECD Countries*. Organisation for Economic Co-operation and Development, Paris.

OECD (1996) *Saving Biological Diversity: Economic Incentives.* Organisation for Economic Co-operation and Development, Paris.

OECD (1997) *Environmental Indicators for Agriculture.* Organisation for Economic Co-operation and Development, Paris.

PARCOM (1993) *Report on Nutrients from Agriculture.* Oslo and Paris Conventions for the Prevention of Marine Pollution – Fifteenth Meeting of the Oslo and Paris Commissions, Berlin (14–19 June).

Reus, J.A., Weckseler, H.J. and Pak, G.A. (1994) *Towards a Future EC Pesticide Policy: An Inventory of Risks of Pesticide Use, Possible Solutions and Policy Instruments.* Centre for Agriculture and Environment, Utrecht.

Robinson, R. (1999) The Environment Protection Viewpoint. *In Accounting for Nutrients: a Challenge for Grassland Farmers in the 21st century* (Ed. A.J. Corrall). Occasional Symposium No. 33. British Grassland Society, pp. 127–136.

Romstad, E., Simonsen, J. and Vatn, A. (Eds) (1997) *Controlling Mineral Emissions in European Agriculture.* CAB International, Wallingford.

Runge, C.F. (1994) The environmental effects of trade in the agricultural sector. In: *The Environmental Effects of Trade.* Organisation for Economic Co-operation and Development, Paris, pp. 19–54.

Scheele, M. (1997) The Decomposition Approach: Spatially Differentiated Analysis and Implementation of Environmental Strategies. In: *Controlling Mineral Emissions in European Agriculture* (Eds. E. Romstad, J. Simonsen, and A. Vatn). CAB International, Wallingford, pp. 41–58.

Tamminga, G. and Wijnands, J. (1991) Animal Waste Problems in the Netherlands. In: *Farming and the Countryside: An Economic Analysis of External Costs and Benefits* (Ed. N. Hanley). CAB International, Wallingford, pp. 117–136.

Taylor, G. and Welford, R. (1994) Environmental strategies of leading UK supermarket chains. In: *Cases in Environmental Management and Business Strategies* (Ed. R. Welford). Pitman, London, pp. 125–135.

Ward, N. (1996) Pesticides, Pollution and Sustainability. In: *The Rural Economy and the British Countryside* (Eds P. Allanson and M. Whitby). Earthscan, London, pp. 40–61.

Waters, G.R. (1994) Government policies for the countryside. *Land Use Policy* **11**, 88–93.

Watson, N., Mitchell, B. and Mulamootil, G. (1996) Integrated resource management: institutional arrangements regarding nitrate pollution in England. *Journal of Environmental Planning & Management* **39**, 45–64.

Webb, J., Loveland, P.J., Chambers, B.J., Mitchell, R. and Garwood, T. (2001) The impact of modern farming practices on soil fertility and quality in England and Wales. *Journal of Agricultural Science (Cambridge)* **137**, 127–138.

Author Index

Subject Index